Modern Robotics: Mechanics, Systems and Control

Modern Robotics: Mechanics, Systems and Control

Julian Evans

Larsen & Keller
www.larsen-keller.com

Modern Robotics: Mechanics, Systems and Control
Julian Evans
ISBN: 978-1-64172-075-5 (Hardback)

📘 Larsen & Keller

Published by Larsen and Keller Education,
5 Penn Plaza,
19th Floor,
New York, NY 10001, USA

Cataloging-in-Publication Data

Modern robotics : mechanics, systems and control / Julian Evans.
    p. cm.
Includes bibliographical references and index.
ISBN 978-1-64172-075-5
1. Robotics. 2. Robots--Kinematics. 3. Robots-- Control systems. 4. Mechanics, Applied. I. Evans, Julian.
TJ211 .M63 2019
629.892--dc23

For more information regarding Larsen and Keller Education and its products, please visit the publisher's website www.larsen-keller.com

Table of Contents

Preface

The field of robotics is involved in the design, construction and use of robots and their control systems. These are developed with the objective of minimizing human effort or for substituting for humans in environments, which are dangerous for human survival such as bomb detection and deactivation, in space, etc. Robotics integrates the techniques of electrical engineering, artificial intelligence, mechanical engineering, etc. for the conception, operation and manufacture of robots. Some common areas where robots are being used include medicine, surgery, military, and manufacturing, where efficiency and precision is of the utmost essence. The components of a robot are a power source, actuators, sensors and manipulators. Solar, nuclear and hydraulic power can be used to drive a robot. This book elucidates the concepts and innovative models around prospective developments in the field of robotics in the modern scenario. Some of the diverse topics covered in this book address the mechanics, systems and control of robotic systems. It aims to serve as a resource guide for students and experts alike and contribute to the growth of the disciplines.

To facilitate a deeper understanding of the contents of this book a short introduction of every chapter is written below:

Chapter 1, The field of robotics is concerned with the construction, design, operation and use of robots and computer systems for information processing, control and feedback. This chapter introduces in brief the principles of robotics. It includes a detailed analysis of its various sub-fields such as biorobotics, cloud robotics, cognitive robotics, developmental robotics, etc. **Chapter 2**, Robots are used in diverse environments and for varied applications. Depending on the potential application, robots can be categorized as military robots, industrial robots, mobile robot, etc. This chapter has been carefully written to provide an overview of common robot types and their applications. **Chapter 3**, Some of the significant components of a robot are the power source, actuator, electric motor, sensors, etc. This chapter explores the diverse components integral to the design of robots that are important for the effective function of a robot. It also discusses in detail about robotic non-destructive functioning, forward chaining, servomechanism, etc. **Chapter 4**, Mechatronics is a field of engineering that unifies the principles and methods of electronics, mechanics, robotics and computing to develop more economical and simpler systems. An important example of a mechatronic system is an industrial robot. This chapter explores the fundamentals of mechatronics, modeling of mechatronic systems and applications of mechatronics. **Chapter 5**, Automation is the technology of performing various operations without human assistance. It can be used for switching on telephone networks, stabilization and steering of aircraft and ships, operating machinery, etc. This chapter investigates the uses of robotics and automation. It includes topics on autonomous car, autonomous logistics, cyborg, laboratory automation, etc.

Chapter 6, The fields of mechatronics and robotics have witnessed significant advancement in the past few decades owing to progress in science and technology. The topics elucidated in this chapter address the varied robotic systems, such as care-O-bot, bowler communication system, etc. as well as discusses about real-time path planning, robot interaction language, perceptual control matter, etc.

Finally, I would like to thank the entire team involved in the inception of this book for their valuable time and contribution. This book would not have been possible without their efforts. I would also like to thank my friends and family for their constant support.

Julian Evans

Introduction to Robotics

The field of robotics is concerned with the construction, design, operation and use of robots and computer systems for information processing, control and feedback. This chapter introduces in brief the principles of robotics. It includes a detailed analysis of its various sub-fields such as biorobotics, cloud robotics, cognitive robotics, developmental robotics, etc.

Robotics is the industry related to the engineering, construction and operation of robots – a broad and diverse field related to many commercial industries and consumer uses. The field of robotics generally involves looking at how any physical constructed technology system can perform a task or play a role in any interface or new technology.

Other than the sci-fi movies, the Robots can be seen around us assembling the cars, in bottling factory etc. The robots have been in the industry for last two decades because of their continuous working ability in an atmosphere, where humans are not even able to stand for minutes, without any supporting equipment, like space.

Since recent years, the researchers are continuously trying to develop more intelligent and self-reliant robots for various applications. The Honda's ASIMO is the best example of such intelligent robots. The human (Homo sapiens) is the most intelligent creature in the world because he has senses like vision, touch etc. The researchers are trying to emulate human capabilities by creating "Robo sapiens" having all these senses.

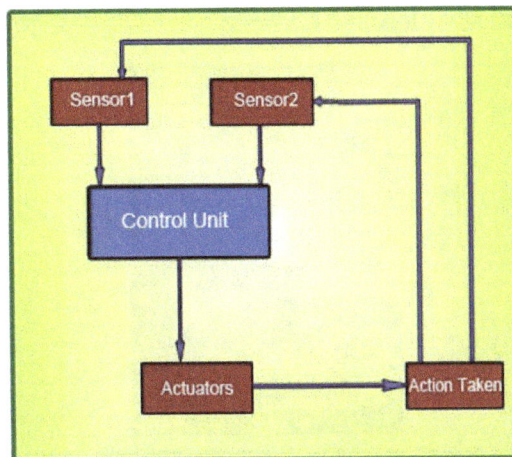

Figure: Simple Block Diagram Showing Different Functions Of Control System In Rototics.

Basic structure of robots is very much similar to humans. How do humans sense? For example a human sees something and sends neural signals to the brain via neurons and reacts accordingly. The development of all these senses artificially is achieved through 'Sensors'. Sensors are the transducers which receive the physical changes of the environment and convert them into electrical or electronic signals. These analog signals are converted into digital by using analog-to-digital convertors. Control system functions as a brain in robotic systems.

PID controller is a popular method for controlling robots. Point of the control system is to get robot actuators do what one wishes to. An encoder on actuator, basically a sensor, determines what all is changing. Program that one writes, defines final results and actuators make changes. Other sensor senses environment, providing robot a better sense of working.

Among different components, actuators are the most important for robotic automation. This component helps in conversion of stored energy in movement. Usually, these are the electric motors, although compressed air and chemical actuators are also available for use. Stepper motors rotates in controllable motions and are commanded by controllers. Ultrasonic motor uses fast vibrating piezo-ceramic element for causing motion. With the help of compressed air, air muscles work exactly like human muscles, contracting and expanding.

Robots are made from an assortment of materials and are driven in many ways too. These can be constructed from sturdy, heavy steel or light weighted plastics. Surgical robots and Robotic arms have uncomplicated rotational joints. These are driven by hydraulics and electrical motors along with longitudinal joints that are moved with rotating screws. Some mobile robots have various wheels that can operate on various planes while others can walk on different terrains with multiple legs. You can find robots with sensing mechanisms like cameras operating as eyes and touch sensors for feeling the environment.

Interesting Robots Developed Lately

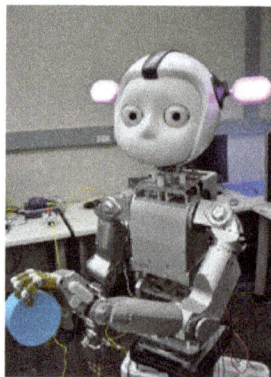

Image of simon robot developed at socially intelligent machines lab at georgia institute of technology that can learn and adapt to humans

Socialization of the robots has been important research area lately. Researchers have been trying to provide robots with the social skills so that they can become much better with assistance at offices, schools, homes and other places. TR35 innovator, Andrea Thomaz invented robots that can grasp tasks from the human instructions with use of verbal instructions, expressions and gestures. University of Carnegie Mellon invented a robot that guides interactions with the help of eye contacts for suggesting it's time for speaking.

Researchers at Californian University invented machine learning programs that allow robotic head to build up better face expressions. By checking itself out in a mirror, this type of robot can study the way in which its body parts move and can develop new expressions.

Types of Robots

From popular droids in the movies Star Wars, C3PO and R2D2, to robot maid in Jetsons, to Schwarzenegger in Cyborg's role in Terminator to Robocop, Matrix, IRobot, humans have always been fascinated by robots. While earlier humans were just fascinated by these robots, now they have them for real. Various types of robots have come into existence.

Industrial robots are getting popular by the day for their improved productivity as compared to humans. These robots are being used in a variety of industrial tasks. Jobs that involve speed, reliability, repetitiveness, and accuracy can be tackled finely by robots. In last 30 years, robots have been handling automatic production lines of automobile industry. Framework of vehicles is transported with the conveyor belt and is affixed, wielded, painted, assembled with the help of various robot stations. Other jobs that industrial robots handle are packaging and palletizing goods, laboratory applications, dispensing jobs, and others in electronic industry.

Industrial Robots

An Image Showing Typical Industrial Robots in Factories

Mobile robots, called Automated Guided Vehicle are being used for transportation of materials in a range of places like container ports, hospitals, warehouses, using markers placed in lasers, floor, and vision. Such robots can perform tasks that are non-repetitive and non-sequential in complex environments. Hence these are also called intelligent robots.

Idea of the robots doing the agriculture work may seem out of fiction book but many robots are in experimental stage of being used for work of ploughing fields, harvesting and gathering. Telerobots on the other hand, are used in the places that prove to be harmful for humans and are inaccessible too. These robots are controlled by human operators at distance. Some of these robots are also used in laparoscopic surgery. These robots being used in nuclear power plant where they tackle hazardous materials. Mobile robots are also proving to be useful in space exploration.

Japanese are using more of service robots. This category involves robots that are used outside industrial facility and can be categorized into ones used for professional and others for personal use. Robots for personal use are getting popular because of their advanced sophistication in the Artificial Intelligence.

Problems and Concerns with Robots

Concerns and fears about the robots have been constantly expressed in a range of films and books. Common theme for the concern is development of master piece that is intelligent and conscious enough to destroy humans. With technology growing at a fast pace, Artificial Intelligence is not a surreal concept. If robots have artificial intelligence, they would make the common sense just like humans.

Though robotics requires programming and instructions, they are prone to errors. Even if there is slightest of inconsistency in programming sub routines, companies using them have to bear huge costs. Right now, the problem robots face is that they lack logic, which takes away sense of recognizing mistakes away from them. One of the biggest limitations is that these robots are limited to commands. If they have to face any situation outside programming, they are of no use.

Future of Robotics

Robots are being used for a range of tasks including the ones that are repetitive, dirty and dangerous for humans. Robots are being used for searching and mapping out areas that are unreachable to humans including other planets and battlefields. On the pace with which robots are progressing, soon there would be more advancement in three main fields including battery size, life and weight, swarm robotics and artificial intelligence.

Future robotics would be long lasting as well smaller. And these robots would prove useful in the tasks of exploration and clear up operation. Right now, the artificial intelligence is very basic in robots but it is expected to get better with ability to act and think in future. This would let them do almost everything that a human can do. Swarm robotics would also be an imperative development in robotics in near future. These robots would control various cooperating robots. Future robots might not need human control for performing various tasks.

Aerial Robotics

The term aerial robotics is often attributed to Robert Michelson, as a way to capture a new class of highly intelligent, small flying machines. However, it is clear that the range of systems and activities covered under the label aerial robotics could extend much further, and that its roots can be found far back in the beginning of the 20th century, together with the birth of aviation. Behind the word aerial robotics we can find several meanings: it could mean robotic flying machines, that is, a mission-independent, platform-oriented concept; however, it could also mean robotics that uses flying machines, that is, a platform-independent, mission-oriented concept. Finally, it could mean a combination of the above that is a description of the robotic platform, together with its robotic mission. In aerospace jargon, robotic flying machines are commonly referred to as unmanned aerial vehicles (UAVs), while the entire infrastructures, systems and, human components required to operate such machines for a given operational goal are often called unmanned aerials systems (UASs). Finally, it is worth noting that many current manned aerial systems definitely carry relevant features of some robotic systems, and so much of this discussion is relevant to manned aircraft.

Applications of Aerial Robotics

Listing all possible applications of aerial robotics is very challenging. However, there are fewer actual implementations, because of the necessity for the corresponding operations to comply with stringent air safety regulations. In the following, a brief description of possible and current applications is provided.

Possible Applications of Aerial Robots

The list of possible applications of aerial robots is long. According to such applications fall within nine categories:

- Remote sensing such as pipeline spotting, power line monitoring, volcanic sampling, mapping, meteorology, geology, and agriculture, as well as unexploded mine detection.

- Disaster response such as chemical sensing, flood monitoring, and wildfire management.

- Surveillance such as law enforcement, traffic monitoring, coastal and maritime patrol, and border patrols.

- Search and rescue in low-density or hard-to-reach areas.

- Transportation including small and large cargo transport, and possibly passenger transport.

- Communications as permanent or ad hoc communication relays for voice and data transmission, as well as broadcast units for television or radio.

- Payload delivery e.g., firefighting or crop dusting.

- Image acquisition for cinematography and real-time entertainment.

Military applications of aerial robots follow the same descriptive lines, with a particular emphasis on remote sensing of humans and critical infrastructure, surveillance of human activity, and payload delivery (bombs, missiles, and ad hoc ground infrastructures devoted to communication and surveillance).

Current Applications

Current applications of aerial robots are somewhat fewer and they are at present driven by the military context.

Aerial Observations

The most important application of aerial robots is aerial observations, which can then be used for terrain map- ping, environmental surveys, crop monitoring, target identification etc. There is a divide, however, between the state of the art for military applications and civilian applications, detailed below.

- Military Operations: Military and government use of aerial robots has sharply increased in recent years in war zones. As a result, dozens of vehicles are now delivered every month, and end up flying in "hot" areas around the globe, most notably in Southwest Asia. The machines being flown range from man-portable machines flying at low altitudes, such as the Pointer or Raven aircraft, to mid-sized machines such as the Aerosonde, Seascan, or Shadow unmanned vehicles, to larger-sized vehicles such as the Predator or Global Hawk. Their wings span from a meter or so for the smaller vehicles to 35 m for Global Hawk (the same as a Boeing 737). Besides their use in military areas or war zones, these machines now find applications in border surveillance, with a particular interest in oceanic borders, where vehicles operate in desert or quasi-desert areas.

- Civilian and Private Applications: Current civilian applications of aerial robots for surveillance and observation remain sporadic and ad hoc: unlike many other robotic devices, civilian aerial robots do not operate in closed environments but in civilian airspace, which is subject to strong safety regulations that do not yet systematically accommodate aerial robots. Consequently, the current trends in civilian aerial robotics are as follows.

Small-scale, intermittent civilian aerial robotic applications tend to happen in relatively isolated environments (e.g., for film making or environmental surveys), and often follow the safety and operations rules most familiar to their operators, derived from model aircraft operations. Most often, the operated machines do in fact bear much resemblance to radio-controlled model airplanes. Other intermittent applications involve the use of unmanned vehicles for specific reconnaissance tasks, such as the detection of fish banks from trawlers. Such a task constituted one of the original purposes for the development of machines such as the Seas can unmanned aerial vehicle.

Long-term scientific applications such as atmospheric sampling experiments to benefit considerably from aerial robots. One report reads:

"From March 6 to March 31 2006, we probed the polluted atmosphere over the North Indian Ocean with lightweight unmanned aerial vehicles (or UAVs) fully equipped with instruments. This UAV campaign launched from the Maldives laid a solid foundation for the use of UAVs to study how human beings are polluting the atmosphere and their impact on climate, including global warming."

Because such activities naturally require much planning ahead, special permits can be obtained from aviation authorities within time limits that do not significantly affect the overall experimental project. Other scientific missions led with success include, where the authors were able to survey Mount St. Helens (then active) by taking advantage of the temporary interdiction to fly in the vicinity of the volcano.

With the progressive introduction of aerial robots in the regulatory framework of many countries, we believe that intermittent applications of aerial robotics in populated areas will eventually become commonplace. However, this requires that flight authorizations be de- livered within a fraction of the time duration of the event: for example, firefighting operations are often triggered within a few seconds of the fire alert. Permits for aerial robotic support should therefore be delivered about as quickly if they are ever to be embraced by firefighters.

Payload Delivery

Under the heading payload delivery, we find the numerous applications of aerial robots aimed at delivering solid, liquid or gaseous products in areas that are hard to reach for humans. So far, the most successful civilian application has been chemical crop spraying using small unmanned helicopters. Leveraging the high costs and prices associated with crop culture in Japan, several thousand helicopters have been purchased by farmers, resulting in a profitable operation both for themselves and for the helicopter manufacturers, among them Yamaha and Yanmar. However, this application remains unique and involved the involvement of Japan's government for it to be successful.

Besides this particular application, military appli- cations form the bulk of unmanned aerial robotics for the purpose of payload delivery, beginning in its crudest form with missiles, and evolving towards cruise missiles, able to navigate for thousands of miles and reach their targets with high precision. One of the most talked-about recent military application of aerial robots for payload delivery involves the Predator aircraft equipped with Hellfire missiles.

General Characteristics of Current Applications

Level of Autonomy

Most, although not all, aerial robots currently under operation are automatically controlled as far as their dynamics are concerned. However, higher levels of au- tonomy, such as path planning, object detection, and recognition and mission management involve human operators, who always remain in contact with the flying machine. Thus not much distinguishes current aerial robots from traditional manned aircraft, except that the pilot sits on the ground rather than in the air.

As such, most of today's operational aerial robots may be justifiably called remotely piloted vehicles (RPVs).

Basic Aerial Robot Flight Concepts

Aerial Robot Flight and the Importance of Scales

Like all flying machines, the performance of aerial robots depends extensively on: (1) their size and (2) the characteristics of their lifting mechanisms (wings, rotors). A detailed description of vehicle flight mechanics is outside of the scope of this chapter; we can, nevertheless, recall a few fundamental and useful notions critical to successful flight. The reference is an excellent and entertaining introduction to the subject, while offer a more academic perspective on the matter.

Flying wing (Northrop's YB47) and its shrunk version flying together

One important quantity is the mass of a flying machine. Roughly speaking, the mass of a flying ma- chine is proportional to its volume, and therefore grows like the cubic power of its size. Another quantity is the lifting forces that keep a vehicle up in the air; these are proportional to the pressure exercised on the lifting surface (rotor or wing), times the area of the lifting surface, that is, roughly the second power of the vehicle size. The pressure itself is proportional to the density of the surrounding atmosphere (it need not be air only, think of Mars), multiplied by the square of the average velocity of the gas molecules relative to the lifting surface.

For illustrative purposes, consider the flying wing shown in and a notional scaled-down version of it flying together. To make matters simpler, we assume that the scaled-down wing is about half the size of the full-sized wing. We now examine the impact of scales on the way these wings must fly.

Consider for example the lift created by the full- scale flying wing depicted in: it is proportional to $S\rho V 2\alpha$, where S is its total surface, ρ is the air density, V is the wing speed relative to the surrounding air, and α is the angle of attack (roughly speaking the angle between the wing chord and the flow of air).

To get an idea of the importance of scales, and following arguments developed in much greater detail in, we now examine the requirements for the scaled-down wing to fly at the same speed as the large wing, assuming all its components are shrunk by a factor two in size as shown in the picture, and examine the consequences of having to meet such requirements.

First, the mass of the wing roughly gets divided by a factor 8. However, its lifting surface has shrunk by a factor of 4 only. So, if we were to fly this smaller wing at the same speed, same altitude, and same angle of attack as its big sister, the total generated lift must be $S/2\rho V 2$, that is, twice as much as necessary to balance out the effect of gravity.

Several solutions to this issue are possible: to reduce the the actual wing dimensions at constant mass, slow it down, or reduce its angle of attack.

Shrink the Wing

To obtain the proper lift (while keeping the speed and angle of attack constant), we must shrink the wing area by another factor of two, or the wing dimensions by a factor $\sqrt{2}$. Thus we already see one important conclusion, which is that, at equal speed and angle of attack, the relative size of the wings with respect to the over- all vehicle must shrink as the overall vehicle size goes down. Borrowing again from, this explains much of why a Boeing 747 looks, with its large deployed wings and relatively narrow fuselage, like a condor while a B737 feels more like a puffin, and the smaller Embraer 145 is like a dart, as shown in figure All these aircraft can fly about the same speeds and altitude ranges.

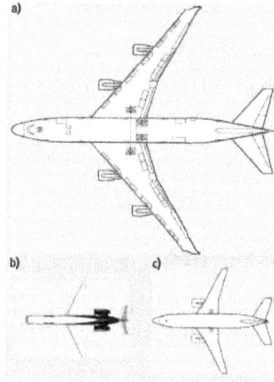

Relative fuselage and wing sizes for various aircraft: (a) Boeing 747:
(b) Embraer 145: (c) Boeing 737

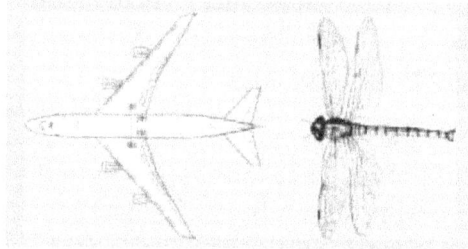

Boeing 747 and dragonfly

Reduce the Speed

If this option is chosen, then speed must be divided by a factor $\sqrt{2}$ for our scaled model to balance lift and weight. The consequences of reducing speed are many: the time required for mission completion of course in- creases. On the other hand, the drag generated by the flying machine (and which must be paid for by the propulsion system) goes down.

Pushed to their limits, the consequences of slowing down the vehicle as it shrinks can be quite dramatic: consider a dragonfly (one of the role models for micro- aerial robots) trying to land next to a Boeing 747 at the same airport. The figure above shows that the two share (very roughly) the same proportions.

For the sake of simplicity, assume that the dragonfly is the 1/1000 scaled-down version of the Boeing 747. In order for both to fly level, and according to our rule, the dragonfly must fly a factor $\sqrt{1000} = 32$ slower than the B747. Assume the B747 flies at 500 km/h; that makes the dragonfly fly at about 17 km/h. Imagine now that the weather is gusty, with winds topping 30 km/h. The 747 (and its passengers) will see little variation in air-speed (from 470 to 530 km/h), and the variation in produced lift will be 33%, enough to shake the aircraft a bit, but not unusually bad. As for the dragonfly, the same gusts will create airspeed variations of well over 100%, and the produced lift will vary from zero to five or six times the nominal lift. A rough ride naturally follows, and indeed, the flight of smaller vehicles often looks much less smooth than that of large ones.

Reduce the Angle of Attack

The latter option, reducing the angle of attack, rests upon the fact that, roughly, the lift created by a wing (or a rotor), is a linear function of the angle of at- tack. This makes it possible to fly about the same speed with a scaled-down model of a flying machine. However, this option comes with significant draw- backs, especially for fixed-wing air-craft. In particular, the sensitivity of the lift created by the wing to external pertur-bations (e.g., air turbulence and wind gusts) would again be higher, creating another recipe for bumpy rides.

The previous considerations about the forces acting on aerial vehicles also apply to mo-ments: consider the flying wings shown in, and assume that their density (mass per unit volume) is constant throughout. Their angular inertia about any axis is proportional to the fifth power of their size. On the other hand, the forces that apply to the wings are proportional to their area; thus when moments are computed, forces are multiplied by distances, and the resulting moments be- come proportional to volume, that is, the third power of vehicle size. Consider then the angular momentum equation where J is the moment of inertia of the vehicle and M is the applied torque.

$$J\ddot{\theta} = M$$

The term to the left of de-creases much faster with vehicle size than that to the right of the equation. As a consequence, we might immediately conclude that the scaled-down flying wing is inherently much more maneuverable than the larger one, in the sense that it can change orientation much faster.

This opens up a wealth of possibilities for robotics: venturing into the world of small flying robots, enabled by improvements of battery power and computation densities opens new possibilities in terms of defining the way these vehicles fly and interact with their environment.

Propulsion Systems

Several propulsion systems exist for aerial robots, including: jet, internal combustion, rocket, and electric. Older but recurrent options also include pulse engines such as those used on the German V1.

Owing to established aircraft and helicopter propulsion technologies, internal combus-tion engines and jet engines form the bulk of the propulsion means for medium to large-sized operational vehicles (50 kg or more), allowing many of them to fly reliably over periods of several hours to several tens of hours. When considering operational robots, the kind of fuel used matters: preference is given to fuels all-ready used in other devices, and preference goes to heavy fuels, which are less prone to sudden and dangerous combustion or explosion, for example after a crash.

(b) Mars aircraft

Electric propulsion systems, once unthinkable, have become a reality for several small-sized aerial robots, thanks to the development of affordable brushless electric engines and lightweight batteries. Initially developed for computer and communication applications, these batteries have been very quickly adopted by small- sized (a few kg) aerial robots such as Aerovironment's Pointer and Raven aircraft, which are able to fly over periods exceeding one hour. The National Aeronautics and Space Administration's (NASA) Pathfinder un- manned aircraft combines lightweight electric engines with wing-mounted solar panels to yield the aircraft shown in figure.

A notable departure from these propulsion systems is the Mars airplane's propulsion system with an inert, low-density atmosphere on Mars, such a vehicle relies on a rocket engine for propulsion.

Flight Vehicle Types and Flight Regimes

Several vehicle types form the bulk of aerial robots, including fixed-wing machines, helicopters, flapping wing systems, and combinations thereof. The boundaries be- tween these vehicle types are, however, mostly inherited from historical developments and intellectual stove- piping, rather than any fundamental guidelines dictated by the laws of mechanics and thermodynamics. For that reason, it is easier and more logical to introduce different *flight regimes* than flight vehicles, although to every regime there naturally corresponds one particular vehicle.

There are essentially two flight regimes. In the first regime, called hover, the speed of the vehicle relative to the surrounding air is small, such that few or no forces act on the vehicle except those resulting from the propulsion system itself. In the second regime, which we may call cruising flight, there is a significant relative speed between the vehicle and its surrounding environment, and significant aerodynamic forces act on the vehicle; these aerodynamic forces then largely dominate those generated by the power system.

Hover

Hover is the condition when the vehicle body does not move significantly with respect to the air mass surrounding it. Under these conditions, only propulsion systems are available to keep the vehicle up in the air (for heavier-than-air systems). Helicopters epitomize these situations, and they are especially designed to sustain such hover conditions over long periods of time. Helicopters come in all sizes and shapes. Robotic helicopters are best represented by Yamaha's R-50, and now RMAX models, and both have been a staple of airborne robotics research for years at several academic institutions because of their reliability and available payload, which allows them to carry many instruments of interest to robotics research (including navigation sensors such as video cameras, laser range finders, and radars). With the evolution of the economic and political context, one is bound to see other such machines abound in the future. Indeed, the ability to hover is extremely useful for delivery/pickup of materials, rescue missions, and, in general, any operations that require close proximity to rugged terrain.

Helicopters are not the only vehicles capable of hover, see for example the hover-capable fixed-wing air- craft in. Hovering aircraft, such as tail sitters have been tested successfully since the 1950s at the very least, and it is a classic trick for experienced remote control pilots to hover airplanes. Transitions from hover to forward flight and back have been automated As radio-control (R/C) equipment shrinks in size and mass, new generations of hovering vehicles will become available. Some of these vehicles include micro air and flapping wing vehicles.

Hovering flight typically not very fuel inefficient: the fuel consumption of a hovering vehicle can exceed that of a fixed-wing vehicle by an order of magnitude of more. This kind of consideration has led manufacturers to seek some of the mixed configurations shown in figure.

Cruising Flight

During cruising flight the aerial robot mostly uses its available surfaces and its speed relative to the surrounding atmosphere to generate lift and maintain altitude. Unlike hovering flight, cruising flight usually results in the aerial robot constantly meeting fresh air, which makes the range of adverse events to flight quite narrower. This, of course, is not true in the case when aerial robots fly in formation, in which case turbulence created by one robot may affect its neighbor(s), sometime adversely, and sometimes positively.

Robotic airplanes epitomize fuel-efficient cruising flight, with large and highly optimized wings. Both aircraft are part of current NASA programs. While optimized for flight, the wings of the Mars aircraft must also be optimized for tight packaging and deployment constraints at the end of its long trip from Earth to Mars.

While many fixed-wing systems are optimized for cruising flight, any system in forward flight operates according to the same principles; for example, a helicopter in forward flight operates like an airplane whose wing a flat disc is spanning the area covered by its rotor. Constant-velocity cruising flight that generates lift from the available aerodynamic surfaces is more fuel efficient than hovering.

Stalled and High-Angle-of-Attack Flight

This flight condition can be seen as a transitional flight condition, where the characteristics of both forward flight and hover are present. Typically, an aircraft stalls when its tries to maintain altitude at low speeds: flying level at lower speeds forces the aircraft's angle-of-attack to increase for the wings to produce more lift. How- ever, past a critical angle of attack, the trend reverse and the lift produced by the wing decreases as the angle of attack keeps increasing. This reduced aero- dynamic lift must then be compensated by increased throttle, resulting in a situation where the aircraft propeller not only acts as a means to move the aircraft forward, but also directly participates in maintaining aircraft altitude.

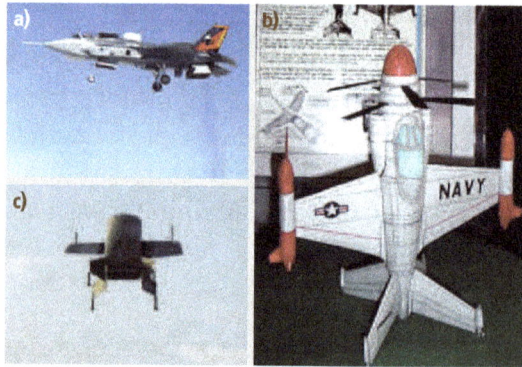

Non-helicopter, hover-capable vehicles: (a) Join Strike Fighter, (b) 1950s tail sitter aircraft and (c) Aurora Flight Sciences' Golden Eye 100

This flight condition, whether experienced on a helicopter or airplane, often results in important changes of the effect of control mechanisms, for example, a stalled Piper Tomahawk trainer aircraft at low throttle setting will experience ineffective ailerons, while its rudder efficiency will shift from yaw axis to roll axis control.

Lighter-than-air Systems

One way to deflect some of the concerns associated with high fuel consumption of heavier-than-air aircraft is to rely on lighter-than-air vehicles. While these vehicles are often associated with spectacular accidents and slow motion, they also offer an unmatched capability to fly for long periods of time (more than 48 h) and to do so silently. Establishing control over lighter-than-air vehicles can be, however, some-what challenging. In particular these vehicles are quite sensitive to winds and often tend to go where the wind takes them. Smaller platforms used for research must therefore evolve in closed environments. The use of such vehicles for outdoor research can be quite daunting, because their large size requires considerable infrastructure to store them.

Anthrobotics

Technology is accelerating at an ever increasing rate. Each year, we develop smaller and smarter systems, systems that allow us to interact with information in ways that previous eras only dreamed about. In fact, given their ability to process, identify, and categorize information—and their uncanny ability to synthesize information and make judgments—many of our systems seem to be developing a true form of intelligence. In this respect, it seems that the dawning age of AI is truly upon us.

It relies on a philosophical view of humans as being the technological animal par excellence. We code and de-code our protocols under the dialectic influence of the creation of the real. Our functionalism can be called collective robotism. Human societies are organic and artificial, and at every moment, as social anthrobots, we're products and producers, partly creators and partly created, partly automata and partly agents capable of adaptability, self-actuation, and sense-making.

Anthrobotics is a working hypothesis towards an interdisciplinary science of world-forming.

Anthrobotics is the choice to consider the human-machine intertwining from the perspective of organized and evolving collectives rather than separated individual entities. Association is not what happens after individuals have been defined with few properties, but what characterizes entities in the first place: this is a conscious step away from methodological individualism. Individual users co-emerge as social agents from the matrix of a social process.

Anthrobotics is a hypothesis that says: let's look at the human-machine intertwining as a dynamic union of more or less institutionalized collectives involved in processes of world forming. The goal is to facilitate the implementation of more plural and harmonious forms of shared natural-artificial forms of live.

It's been said that the relationship between humans and technology can at times resemble an infinite game where the principal outcome is to continue playing.

Anthrobots that pre-date the computer age, such as institutions, organizations, corporations, nation states, rituals, collective organized projects, etc. provide blueprints that we can use as models for understanding and developing more plural and harmonious socio-technical systems.

Looking at pre-computerized groups and their social protocols, where individuals assume, embody, and express different forms of belongingness or esprit de corps could perhaps provide guiding principles for the design of socially embedded robotics. Anthrobotics is not only a matter of social engineering and ethics, but also of policy. If a human collective is an axiomatic, intrinsically normative system, we can aim at more healthy, co-creative, and virtuous systems that favours respectful collaborations within socio-technical assemblages.

It's been said that the relationship between humans and technology can at times resemble an infinite game where the principal outcome is to continue playing. We get caught up in the game, but we can also step back, and remember that our sociocultural games are indeed made-up. By whom? Our worlds are, in one way or another, co-created by us, both by our courage and our cowardice. Our worlds are our anthrobots. We should extend the collective co-creation of these social machines, as proposed for example by philosophers Deleuze and Guattari, who defined human communities as "desiring machines", which is another definition of anthrobotics.

BEAM Robotics

BEAM robots are a type of robot that do not use computers. They are typically cheap to make and can be built within a few days—unlike computer-based robots that can be costly, complex and take years to build.

Born out of the mind of Mark Tilden, a former researcher at the Los Alamos National Laboratory, BEAM robots can be either simple machines consisting of a solar cell, motor, transistors and capacitors or as complex as an 8-jointed, 4-legged walking spider machine.

BEAM (Biology, Electronics, Aesthetics, Mechanics) robots generally do not feature a microprocessor which means they are usually not as flexible as other robots. However, this simplicity makes them robust and efficient at performing the tasks they are designed to do.

The maker community is using BEAM robots, also known as junkbots, to create new ways of building robots. The technology is also a good way for kids to learn about robotics and coding.

With a solar cell on top, this robots moves when light hits it and also has LEDs to light it up.

These robots provide an especially good platform for learning these skills because of their low cost. Also, most BEAM robots use parts that can be found either in recycled electronics such as microwave ovens, portable audio players, or pagers, or that can be easily found through online distributors and kits. The average age of people building BEAM robots is 13 to 17 years; the maker community also includes older electrical engineers and hobbyists.

Types of BEAM Robots that can be Built

The most common types of BEAM robots include:

- Solaroller: A small wheeled robot that collects energy from a solar cell and then bursts forward a few millimeters to a few feet depending on the construction.

- Photovore: Basically two solarollers stuck together to give the robot a phototropic behavior: it moves toward the brightest light source.

- Walker: This is the most common type of BEAM robot, typically containing four legs and two motors. More complex walkers can be built using six legs and up to twelve motors. These types of robots are more functional and can do things such as interplanetary exploration scan for land mines and more.

- Sitters: Immobile robots that can either be used for a light show or to transmit a signal for other BEAM bots to use.

- Squirmers: Stationary robots that use magnetic fields to move, move a display around, vibrate or pivot about.

- Jumpers: Bots that propel themselves off the ground as a means of motion.

- Rollers: A type of solaroller that uses a single motor to drive one or more wheels.

- Swimmers: BEAM robots that move below or on top of the surface of a liquid.

- Fliers: Robots that use a powered rotor for both lifting and propulsion.

With the advancement of technology, more uses are being added to BEAM robots such as using light emitting diodes (LEDs) to light up robots or using developer boards such as Raspberry Pi to give additional functionality.

Behavior-based Robotics

The concept of Behavior-based robotics or BBR was introduced in the mid-1980s, and was championed by Rodney Brooks and others. Nowadays, the behavior-based approach is used by researchers worldwide, and it is often strongly influenced by ethology.

Comparison with Classical AI

BBR approaches intelligence in a way that is very different from the classical AI

Wapproach, and a schematic illustration of the difference is given. In classical AI, the flow of information is as shown in the left panel of the figure. First, the sensors of the robot sense the environment. Next, a (usually very complex) world model is built, and the robot reasons about the effects of various actions within the framework of this world model, before finally deciding upon an action, which is executed in the real world.

Now, this procedure is very different from the distributed form of computation found in the brains of biological organisms, and, above all, it is generally very slow, strongly reducing its survival value. Clearly, this is not the way most biological organisms function. As a good counterexample, consider the evasive maneuvers displayed by noctuid moths, as they attempt to escape from a pursuing predator (e.g. a bat). A possible way of achieving evasive behavior would be to build a model of the world, considering many different bat trajectories, and calculating the appropriate response. However, even if the brain of the moth were capable of such a feat (it is not), it would most likely find itself eaten before deciding what to do. Instead moths use a much simpler procedure.

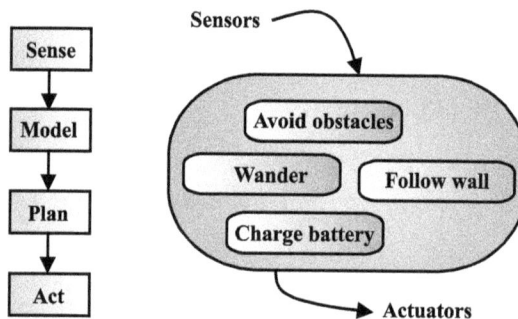

A comparison of the information flow in classical AI (left panel) and in BBR (right panel).
For BBR, any number of behaviors may be involved, and the figure only shows
an example involving four behaviors.

As is evident from the left panel of figure, classical AI is strongly focused on high-level reasoning, i.e. an advanced cognitive procedure displayed in humans and, perhaps, some other mammals. Attempting to emulate such complex biological systems has proven simply to be too complex as a starting- point for research in robotics: Classical AI has had great success in many of the subfields it has spawned (e.g. pattern recognition, path planning etc.), but has made little progress toward the goal of generating truly intelligent ma- chines, capable of autonomous operation.

BBR, illustrated in the right panel of figure is an alternative to classical AI, in which intelligent behavior is built in a bottom-up fashion, starting from simple behaviors, many of which may be running simultaneously in a given robotic brain, giving suggestions concerning which actions the robot ought to take.

Returning to the example of the moth, its evasive behavior is very simple indeed, and is in fact based on only a few neurons and an ingenious positioning of the ears on its body. This simple system enables the moth to fly away from an approaching bat and, if it is unable to shake off the pursuer, start to fly erratically (and finally dropping toward the ground) to confuse the predator.

Behaviors and Actions

An attempt should be made to define the concepts of behaviors and actions, since they are used somewhat differently by different authors. Here, a behavior will be defined simply as a sequence of actions performed in order to achieve some goal. Thus, for example, an obstacle avoidance behavior may consist of the actions of stopping, turning, and starting to move again in a different direction. Note, however, that the definition is not very strict, and that the terms behavior and action remain somewhat blurred.

Intelligent Behavior and Reasoning

The example with the moth shows that intelligent behavior does not require reasoning, and in BBR one generally uses a more generous definition of intelligent behavior than that implicity used in AI. Thus, in BBR, one may de- fine intelligent behavior as the ability to survive, and to strive to reach other goals, in an unstructured environment. This definition is more in tune with the fact that most biological organisms are capable of highly intelligent behavior in the environment for which they have been designed, even though they may fail quite badly in novel environments (as illustrated by the failure of e.g. a fly caught in front of a window). An unstructured environment is an environment that changes rapidly and unexpectedly, so that it is impossible to rely on pre–defined maps, something that holds true for most real-world environment. For example, if it is the task of a robot to transport equipment in a hospital, it must first be able to avoid (moving) obstacles. Putting a complete plan of the hospital into the robot will not be sufficient, since the environment changes on a continuous basis. The robot may pass an empty room, only to find the room full of people or equipment when it returns.

Features of Behavior-based Robots

Different behavior-based robots generally share certain basic features, even though not all features are present in all such robots. To start with, behavior- based robots are first provided with the most basic behaviors such as obstacle avoidance or battery charging, needed to function in an unstructured environment. More complex behaviors are added at a later stage. Second, several behaviors are generally in operation simultaneously, and the final action taken by the robot represents either a choice between the suggestions given by the various behaviors, or an average action formed from the actions suggested by several behaviors. Third, BBR is mostly concerned with autonomous robots, i.e. robots that are able to move freely and without direct human supervision. Finally, the concept of situatedness is a central tenet of BBR: Behavior-based robots do not build complex, abstract world models. Instead, behavior-based robots are generally situated (i.e. operate in the real world), and many behaviors in BBR are reactive, i.e. have a direct coupling between sensors and actuators. However, internal states, which provide the robot with, for example, motivation or short-term memory, are of course allowed and are often very useful, but, to the extent that such states are used, they are not in the form of abstract world models.

Note that, in BBR, it is often necessary to use simulations before an implementation in an actual robot is attempted. While the use of simulations represents a step away from the requirement of situatedness, the simulations used in BBR differ strongly from the construction of abstract world models: Just like a physical behavior-based robot, a simulated robot in BBR will rely only on the information it can gather through its simulated sensors, and the reading of a simulated sensor, in itself, never provides information that the physical counterpart to the sensor would be unable to provide. However, no simulation can capture all the facets of reality. Thus, it is important to test, in real robots, any results obtained through simulations.

In general, the brain of a behavior-based robot is built from a set of basic behaviors, known as the behavioral repertoire. The construction of such a brain can be considered a two-stage process: First the individual behaviors must be defined (or evolved). Next, a system for selecting which behavior(s) to use in any given situation must be constructed as well. Clearly, in any robot intended for complex applications, the behavioral selection system is just as important as the individual behaviors themselves. For example, returning to the example of the hospital robot, it is clear that if it starts to run out of power, it must reprioritize its goals and quickly try to find a power supply, even if, by doing so, it comes no closer to achieving its task of delivering objects.

Bio-inspired Robotics

Biologically-Inspired Robotics is part of a body of work at the intersection of biology and robotics.

As mentioned above, biorobotics actually consists of two different endeavors, biologically-inspired robotics and biorobotic modeling. However, these two endeavors have distinct motivations and criteria for success. Thus, although it is certainly possible for a single research project to make both technological and scientific contributions, it is important to carefully distinguish which aspects of a biorobotic research project are intended as contributions to which field.

In biologically-inspired robotics, the primary goal is technological: Biologically-inspired roboticists wish to build better robots. They look to biology for inspiration because, compared to current robots, the behavior of animals is extremely flexible and robust in the face of environmental contingencies. The hope is that adopting some of the design principles of animals will endow robots with similar flexibility and robustness.

Biological inspiration can be drawn from many aspects of animals, including their behavioral strategies, the physical design of their bodies, and the organization of their nervous systems. Many degrees of biological inspiration are also possible, from vague resemblance to strict emulation. Major issues that must be considered include the degree of realism necessary to reap the benefits of biological inspiration and the separation of incidental biological details from those essential to performance of the task of interest. Often, a roboticist will use ideas from biology as a springboard for new engineering designs, subsequently ignoring biological realism. This is as it should be since, as a technological endeavor, the success of a biologically-inspired robotics project must not be judged by its faithfulness to the biological data. Rather, it must be judged by the extent to which the performance of the biologically-inspired robot improves upon existing technological approaches using whatever performance metrics are standard for that technology.

In contrast, the primary goal of biorobotic modeling is scientific: Biorobotic modelers wish to understand the mechanisms of animal behavior. As such, like any other biological model, biorobotic models must be judged by the extent to which they account for and illuminate the observed behavioral and neurobiological data, as well as the extent to which they generate testable experimental hypotheses. It is becoming increasingly clear that the mechanisms of animal behavior must be sought not only in that animal's nervous system, but also in its body and environment and the dynamics of interaction between these three components (Chiel and Beer). The special advantage of biorobotic models over computational ones derives from the fact that accurately modeling the physical body and world of an animal can be extremely difficult and computationally expensive, whereas the physics comes "for free" in a physically-instantiated biorobotic model. However, a major difficulty in robotic modeling is ensuring that the relevant physical, sensory and motor properties of the robot sufficiently match those of the animal relative to the biological question of interest, for if they do not, then the robot might actually work against the model's biological relevance. Webb provides an in-depth discussion of these and other issues related to biorobotic modeling.

Examples of Bio-inspired Robotics

The Octobot from Harvard University

The Octobot excels in two distinct ways. First, it is a soft robot, replacing all mechanical components with analogous soft systems, and second, it's autonomous. The robot is 3D printed, inlaid with channels for power and movement control, and the movement is powered through pneumatic controls by gas from hydrogen peroxide, which also is the liquid fuel for the robot. A circuit, a soft analog of a simple electronic oscillator, controls when hydrogen peroxide decomposes to gas to inflate the robot. The gas pushes through the limbs and the microfluidic network shuts off corresponding limbs based on external feedback. As one limb starts to deflate, the gas is redirected to another so that the robot can move. The research was headed by Robert Wood and Jennifer Lewis of the Wyss Institute for Biologically Inspired Engineering at Harvard University. Wood states that the "research demonstrates that we can easily manufacture the key components of a simple, entirely soft robot, which lays the foundation for more complex designs".

Hexa: The Six-Legged Robot

Hexa from Vincross is a new six-legged robot that allows users to control it from their smartphone. The robot uses a variety of sensors to determine its orientation and navigate around terrain. It has the ability to scale steps and balance on uneven ground without having the need to control each individual leg. The six-legged design allows the robot to save energy by having better balance capability. According to Andy Xu, COO of Vincross, the robot "only needs three legs to stand on the ground, and we can use the other three legs to maintain balance or climb stairs." The sensors in the Hexa bot are a digital camera with night vision, a 3-axis accelerometer, distance-measuring sensor, and an infrared transmitter. The programming language for the robot is open-source, allowing multiple users to experiment with the robot and fostering a programming marketplace. Vincross hopes to launch the robot for multiple uses as the user community grows, including exploration of dangerous situations like collapsed buildings.

Pleurobot The Robo-Salamander

At the École Polytechnique Fédérale de Lausanne, engineers have designed a robot that mimics the motion of a salamander. The robot imitates the ambulation of the salamander with a unique vertebrate that allows the robot to slither in and out of the water. A salamander in nature can shift from a crawl to a walk to a swim by performing the same motion at different speeds. This appealed to the engineers because one doesn't need to create different mechanisms to achieve different movements, but rather find the optimal mechanism that can perform several movements. The skeleton of the Pleurobot has only 11 spinal segments, down from the original 40 planned segments, which were not critical for the bot's movement. The joints also have reduced freedom of movement. Auke Ijspeert, leader of the project, explains that understanding the neuro-prosthetics of the salamander is important for understanding the human spinal cord and brain interaction. "Being able to re-stimulate those circuits in humans in the long term is something very important," he says, "and for that you need to understand how the spinal cord works".

The Snakebot

Researchers at Carnegie Mellon University have been programming snake robots for years to get them to crawl through rubble and around obstacles. The inspiration of these robots came from observing sidewinders. Their motion can best be described in two terms: vertical and horizontal body waves. Changing the phase and amplitude of these waves allows snakes to achieve enhanced movement. The modular snake robot shown above was designed to pass horizontal and vertical waves in a 3D space. The robot is 2 inches in diameter and 37 inches long, consisting of 16 joints. Each joint is perpendicular to the previous one, allowing it to achieve numerous configurations and a variety of gaits. NASA is also using these snake robots for space exploration possibilities.

Cassie the Bipedal Bot

From Agility Robotics, Cassie is bipedal robot based specifically on the anatomy of ostriches. Cassie has three degrees of freedom at its hips and ankles, and knees that flex in one direction. The manner of walking comes very naturally to the robot due to its design, which also results in a steering mechanism that is similar to humans. The robot is lightweight and is designed to absorb shock just as humans do when they walk. The human-like movements let the robot traverse areas that humans can and its slip- and stumble-resistant feet help provide stability on uneven ground. Cassie's possible jobs can range from delivering products to your door to search-and-rescue efforts.

Biorobotics

Biorobotics is the use of biological characteristics in living organisms as the knowledge base for developing new robot designs. The term can also refer to the use of biological specimens as functional robot components. Biorobotics intersects the fields of cybernetics, bionics, biology, physiology, and genetic engineering.

Biorobotic technologies are often utilized to provide assistance to accommodate a deficiency—either as fully-functioning robots or highly advanced prosthetics; the latter represents one area in which neural engineering and biorobotics intersect as both disciplines are required in order to first signal and then generate movement. Such devices may also be used to measure the state of disease, track progress or offer interactive training experiences that can speed recovery from an injury or stroke.

Biorobotics encompasses a diverse array of disciplines with a myriad of applications. Researchers in Italy, for example, are developing artificial sensing skin that can detect pressure as contact is made with an object. Tactile sensors are important not only for self-standing robots and limb prostheses but as a means of restoring the sense of touch to diabetics with peripheral neuropathy by mimicking sensations normally gleaned by fingerpads and feet. This one application of biorobotics requires contributions from biomedical engineers studying tissue engineering, neural engineering, biomimetics and BioMEMS.

Scientists are also exploring the potential for early diagnosis of autism by monitoring sensory-motor development through mechatronic-sensorized toys, such as rattles with force and contact sensors.

Biorobotics is being used to help train surgeons and dentists using virtual environments that speed the learning process by facilitating epiphanies, or "Aha" moments. It is also being used to assist in actual surgeries, allowing for more precise and less invasive interventions. Endoscopic robots at the tip of a probe can, for example, remove a polyp during a colonoscopy. And mechatronic handheld tools allow surgeons to manipulate their hands at the macro level while affecting similar responses from a mechanical device operating at the micro level. One day, this could even lead to "cellular surgery".

Key to surgical robotics is the sense of touch, or haptics. Many researchers are exploring how to enhance haptic perception and feedback to allow a surgeon to virtually palpate and squeeze tissue and sense how deep to make an incision.

As robots become more sophisticated and embedded in our lives, Human-Robot Interaction & Coordination (HRI&C) has emerged as a sub-discipline that focuses on the behavior and place of robots in society.

Cloud Robotics

Cloud robotics is an emerging field of robotics ingrained in cloud computing. It allows robots to benefit from the powerful computational and communications resources of modern data centers. When computational or storage demands exceed the on-board capacity of a robot, where the massive resources of a data center can supplement their limited local resource.

So cloud robotics includes other internet related aspects of robotics such as trends towards online sharing of open source hardware and software, crowd sourcing of robotics funding tele presence and human based computation.

Cloud-enabled robots could offload CPU-heavy tasks to remote servers, relying on smaller and less power-hungry onboard computers. Using the cloud, a robot could improve capabilities such as speech recognition, language translation, path planning, and 3D mapping.

Advantages of using Cloud Computing in Robotics

1. Cloud based robotics provides shared knowledge, database on real time. This makes robots smarter.

2. Cloud robotics can transfer heavy computer task to the cloud. This leads to cheaper lighter and easy hardware maintenance. Hardware and software updates can be in real time without losing the data.

3. Any old robot which is full on rust and dust can be reuse properly by using hardware and software of cloud infrastructure.

Disadvantages of using Cloud Computing in Robotics

Environmental security – The concentration of computing resources and users in a cloud computing environment also represents a concentration of security threats. Because of their size and significance, cloud environments are often targeted by virtual machines and bot malware, brute force attacks, and other attacks.

Data privacy and security – Hosting confidential data with cloud service providers involves the transfer of a considerable amount of an organisation's control over data security to the provider. For example, every cloud contains a huge information from the clients include personal data. Another problem is once a robot is hacked and controlled by someone else, which may put the user in danger.

Effects of Cloud Robotics on Safety

Autonomous cars can use the cloud data as they navigate accessing voluminous amounts of navigation data. The Google self-driving car already uses maps and images collected by satellites and stored on the cloud.

The problem is that hackers can also access "vast amounts of computing power" and data. Safety systems in factory robots could be maliciously overridden and cause personal injury or even loss of life.

Cars could be provided erroneous data from the cloud causing crashes that could result in the loss of life. Imagine a hacker with access to cloud computing power.

This is where exida can provide value by performing Safety and Security audits. Safety and security are very much codependent for a complete safety solution. Outcomes of the audits lay the groundwork for a robust security and safety architecture.

Examples of Cloud Robotics Project

ASORO labs: Researchers at Singapore ASORO labs have built a cloud computing infrastructure to generate 3D model of environment which allows robots to perform simultaneous localization and mapping. This process is much faster than their computers

LAAS: here scientist are developing robotic objective database for robots to simplify the manipulation task like simple process of opening doors.

Gostai: this French robotic firm has developed a cloud based robotic infrastructure known as GostaiNet which allows robots to perform speech recognition , face detection and other task remotely. Gostai's robot uses the cloud for video recording and voice synthesis.

Google's self-driving cars are one type of cloud-connected robot. The autonomous cars access data from Google Maps and images stored in the cloud to recognize their surroundings. They also gather information about road and traffic conditions and send that information back to the cloud.

Rapyuta is an open source cloud robotics framework based on RoboEarth Engine developed by the robotics researcher at ETHZ.

C2RO (C2RO Cloud Robotics) is a platform that processes real-time applications such as collision avoidance and object recognition in the cloud. Previously, high latency times prevented these applications from being processed in the cloud thus requiring on-system computational hardware (e.g. Graphics Processing Unit or GPU).

Cloud Robotics is still very new and has to undergo a long series of designing and developing. There are many problems to deal with before it goes to mainstream.

Cognitive Robotics

Scope

There is growing need for robots that can interact safely with people in everyday situations. These robots have to be able to anticipate the effects of their own actions as well as the actions and needs of the people around them.

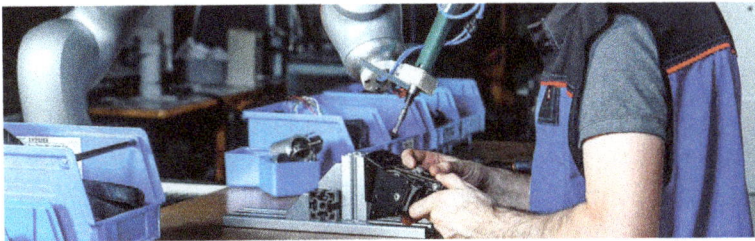

To achieve this, two streams of research need to merge, one concerned with physical

systems specifically designed to interact with unconstrained environments and another focusing on control architectures that explicitly take into account the need to acquire and use experience.

The merging of these two areas has brought about the field of Cognitive Robotics. This is a multi-disciplinary science that draws on research in adaptive robotics as well as cognitive science and artificial intelligence, and often exploits models based on biological cognition.

Cognitive robots achieve their goals by perceiving their environment, paying attention to the events that matter, planning what to do, anticipating the outcome of their actions and the actions of other agents, and learning from the resultant interaction. They deal with the inherent uncertainty of natural environments by continually learning, reasoning, and sharing their knowledge.

A key feature of cognitive robotics is its focus on predictive capabilities to augment immediate sensory-motor experience. Being able to view the world from someone else's perspective, a cognitive robot can anticipate that person's intended actions and needs. This applies both during direct interaction (e.g. a robot assisting a surgeon in theatre) and indirect interaction (e.g. a robot stacking shelves in a busy supermarket).

In cognitive robotics, the robot body is more than just a vehicle for physical manipulation or locomotion: it is a component of the cognitive process. Thus, cognitive robotics is a form of embodied cognition which exploits the robot's physical morphology, kinematics, and dynamics, as well as the environment in which it is operating, to achieve its key characteristic of adaptive anticipatory interaction.

Robot Control

To perform as per the program instructions, the joint movements an industrial robot must accurately be controlled. Micro-processor-based controllers are used to control the robots. Different types of control that are being used in robotics are given as follows:

Point to Point Control Robot

The PTP robot is capable of moving from one point to another point. The locations are recorded in the control memory.

PTP robots do not control the path to get from one point to the next point. Common applications include:

- Component insertion
- Spot welding
- hole drilling
- Machine loading and unloading
- Assembly operations

Continuous-path Control Robot

The CP robot is capable of performing movements along the controlled path. With CP from one control, the robot can stop at any specified point along the controlled path.

All the points along the path must be stored explicitly in the robot's control memory. Applications Straight-line motion is the simplest example for this type of robot.

Some continuous-path controlled robots also have the capability to follow a smooth curve path that has been defined by the programmer.

In such cases the programmer manually moves the robot arm through the desired path and the controller unit stores a large number of individual point locations along the path in memory (teach-in).

Typical applications include:

- Spray painting
- Finishing
- Gluing
- Arc welding operations

Controlled-path Robot

In controlled-path robots, the control equipment can generate paths of different geometry such as straight lines, circles, and interpolated curves with a high degree of accuracy.

Good accuracy can be obtained at any point along the specified path.

Only the start and finish points and the path definition function must be stored in the robot's control memory.

It is important to mention that all controlled-path robots have a servo capability to correct their path.

Stop-to-Stop

- It is open loop system;
- Position and velocity unknown to controller;
- On/off commands stored as valve states;
- End travel set by mechanical.

Developmental Robotics

Developmental robotics (also known as epigenetic robotics or ontogenetic robotics) is a highly interdisciplinary subfield of robotics in which ideas from artificial intelligence, developmental psychology, neuroscience, and dynamical systems theory play a pivotal role in motivating the research. The main goal of developmental robotics is to model the development of increasingly complex cognitive processes in natural and artificial systems and to understand how such processes emerge through physical and social interaction. Robots are typically employed as testing platforms for theoretical models of the emergence and development of action and cognition – the rationale being that if a model is instantiated in a system embedded in the real world, a great deal can be learned about its strengths and potential flaws. Unlike evolutionary robotics which operates on phylogenetic time scales and populations of many individuals, developmental robotics capitalizes on "short" (ontogenetic) time scales and single individuals (or small groups of individuals).

Human intelligence is acquired through a prolonged period of maturation and growth during which a single fertilized egg first turns into an embryo, then grows into a newborn baby, and eventually becomes an adult individual – which, typically before growing old and dying, reproduces. The processes underlying developmental changes are inherently robust and flexible as demonstrated by the amazing ability of biological organisms to devise adaptive strategies and solutions to cope with environmental changes and guarantee their survival. Because evolution has selected development as the process through which to realize some of the highest known forms of intelligence, it is reasonable to assume that development is mechanistically crucial to emulate such intelligence in machines and other human-made artifacts.

Aspects and Areas of Interest

iCub

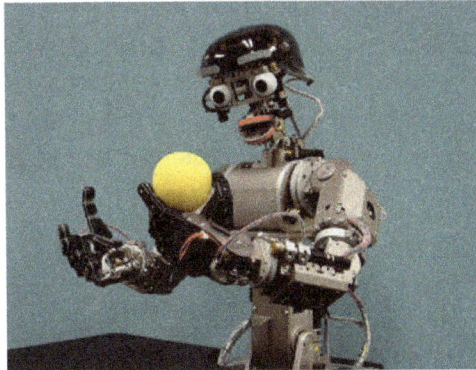

Infanoid

Developmental robotics differs from traditional robotics and artificial intelligence in at least two crucial aspects. First, there is a strong emphasis on body structure and environment as causal elements in the emergence of organized behavior and cognition requiring their explicit inclusion in models of emergence and development of cognition). Although some researchers use simulated environments and computational models, more often developmental robots are embedded in the real world as physical analogues of real organisms. Second, the idea is to realize artificial cognitive systems not by simply programming them (e.g. to solve a specific task), but rather by initiating and maintaining a developmental process during which the systems interact with their physical environments (i.e. through their bodies, tools, or other artifacts), as well as with their social environments (i.e. with people, other robots, or simulated agents) – cognition, after all, is the result of a process of self-organization (spontaneous emergence of order) and co-development between a developing organism and its surrounding environment. Andy Clark uses the term "cognitive incrementalism" to denote the bootstrapping of intelligence, the rationale being that throughout life "you get indeed get full-blown, human cognition by gradually adding bells and whistles to basic strategies of relating to the present at hand". In other words, incrementalism designates the process of starting with a minimal set of functions and building increasingly more functionality in a step by step manner on top of structures that are already present in the system.

The spectrum of developmental robotics research can be roughly segmented into four primary areas of interest. The borders of these categories are not as clearly defined as this classification may suggest and instances may exist that fall into two or more of these categories. We do hope, however, that the suggested grouping provides at least some order in the large spectrum of issues addressed by developmental roboticists.

- Socially oriented interaction: This category comprises research on robots that communicate or learn particular skills via social interaction with humans or with other robots. Examples include research on imitation learning, communication and language acquisition, attention sharing, turn-taking behavior, and social regulation.

- Non-social interaction: These studies are characterized by a direct and strong coupling between sensor and motor processes and the local environment (e.g. inanimate objects), but do not involve any interaction with other robots or humans. Examples are visually-guided grasping and manipulation, tool-use, perceptual categorization, and navigation.

- Agent-centered sensorimotor control: In these studies, the goal is to investigate the exploration of bodily capabilities, changes of morphology (e.g. perceptual acuity, or strength of the effectors) and their effects on motor skill acquisition, self-supervised learning schemes not specifically linked to any functional goal, and models of emotion. Examples include self-exploration, categorization of motor patterns, motor babbling, and learning to walk or crawl.

- Mechanisms and principles: This category embraces research on mechanisms or processes thought to increase the adaptivity of a behaving system. Many examples exist: developmental and neural plasticity, mirror neurons, motivation, freezing and freeing of degrees of freedom, and synergies; research into the characterization of complexity and emergence, as well as the effects of adaptation and growth; practical work on body construction or development. Further work in this area of interest relates to design principles for developmental systems.

Design Principles

By contrast to traditional subjects such as physics or mathematics, which are described by basic axioms, the fundamental ("universal") principles governing the dynamics of developmental systems are unknown. Could there be laws governing developmental systems? Could there be a theory? Although various attempts have been initiated Brooks et al., it is fair to say that to date no such theory has emerged. Towards such a theory, one attractive possibility is to point out a set of candidate design principles. Such principles can be abstracted from biological systems (e.g. they can be revealed by observations of human and animal development), and their inspiration can take place at several levels, ranging from a "faithful" replication of biological mechanisms to a rather generic implementation of biological principles leaving room for other dynamics that are intrinsic to artifacts but are not found in natural systems. It is generally believed that such a principled approach is preferable for constructing intelligent autonomous systems with desired properties because it allows the capturing of design ideas and heuristics in a concise and pertinent way, avoiding blind trial-and-error (for additional information and principles refer to.

Challenges

The further success of developmental robotics will depend on the extent to which theorists and experimentalists will be able to identify universal principles spanning the multiple levels at which developmental systems operate. In what follows, we briefly indicate some of the "hot" issues that will need to be tackled in the future.

- Semiotics: It is necessary to address the issue of how developmental robots (and embodied agents in general) can give meaning to symbols and construct semiotic systems. A promising approach – explored under the label of "semiotic dynamics" – is that such semiotic systems and the associated information structure are not static, but are continuously invented and negotiated by groups of people or agents which use them for communication and information organization.

- Core knowledge: An organism cannot develop without some built-in ability. If all abilities are built in, however, the organism does not develop either. It will therefore be important to understand with what sort of core knowledge and explorative behaviors a developmental system has to be endowed so that it can begin developing novel skills on its own. One of the greatest challenges will be to identify what those core abilities are and how they interact during development in building basic skills.

- Core motives: It is necessary to conduct research on general capacities such as creativity, curiosity, motivations, action selection, and prediction (i.e. the ability to foresee consequence of actions). Ideally, no tasks should be pre-specified to the robot, which should only be provided with an internal abstract reward function, some core knowledge, and a set of basic motivational (or emotional) "drives" that could push it to continuously master new know-how and skills.

- Self-exploration: Another important challenge is the one of continuous self-programming and self-modeling. Control theory assumes that target values and statuses are initially provided by the system's designer, whereas in biology, such targets are created and revised continuously by the system itself. Such spontaneous "self-determined evolution" or "autonomous development" is beyond the scope of current control theory and needs to be tackled in future research.

- Active learning: In a natural setting, no teacher can possibly provide a detailed learning signal and sufficient training data. Mechanisms will have to be created for the developing agent to collect relevant learning material on its own and for learning to take place in an "ecological context" (i.e. with respect to the environment). One significant future avenue will be to endow systems with the possibility to recognize progressively longer chains of cause and effect.

- Growth: As mentioned in the introduction, intelligence is acquired through a process of self-assembly, growth, and maturation. It will be important to study how physical growth, change of shape and body composition, as well as material properties of sensors and actuators affect and guide the emergence and development of cognition and action. This will allow connecting developmental robotics to computational developmental biology.

Microrobotics

Two major discoveries in the world had shaped the future of MICROROBOTICS, the Microscope and the Integrated Circuits, one showed us the micro scale and the other how to work on it. MICROROBOTICS is a field that is getting a lot of attention today, not only from engineers specializing in the field, but also from other fields of engineering, scientists, hobbyists and even the general public. As time has passed, these robots have started off at sizes in millimetres, and have now reached sizes in the micro and nano scale of measurement. Miniaturization of technology has allowed us to push further for wider boundaries towards building smaller as well as more efficient versions of these, and it has proved that small things matter. As these microrobots become smaller and smaller, we discover more ways of applying microrobotics in more and more fields of matter.

Miniaturized robotic systems that make use of micro technologies are termed as microrobots. A microrobot may also be defined as one that possesses traits of a robot in the macro world and has some form of reprogrammable behavior and is capable of adapting, the only difference to a macrorobot being the scale at which they are placed. The terms microrobots or microrobotics are also linked to robots that are able to handle objects and carry operations at the micrometer range.

Microrobots due to their small size can have limited functionality and thus have to work in large groups or swarms in order to sense and affect the environment. The use micro and nanotechnologies in robotics not only reduce the size of robots but also results in reductions of the required recourses and lead to better performance of the robots.

The most challenging aspect in the development of microrobots is the fabrication of micro actuators and micro sensors which can give high efficiency and high stability. To overcome such problems scientists and researchers are combining technologies such as Micro/Nano Electro Mechanical Systems (MEMS, NEMS), nanotechnology and biotechnology.

The miniaturized size and integrations of devices and systems into one small system has the following advantages:

- Reduction in the required resources (mass, volume, power etc.).

- Many discrete devices and components can be replaced by Micro-electro mechanical systems (MEMS) and Nano electro mechanical systems (NEMS).

- The micro robotic systems are usually manufactured in "batch" process and this mass fabrication results in redundancy of critical parts in order to achieve a higher reliability during operation.

The overall cost of the whole systems can be reduced or a given overall cost increases performance with microrobotics.

Microfabrication and Assembly

Fabricating structures on a micro scale is the key for this field of robotics. With inspiration taken from MEMS, Bulk and Surface Micromachining are the major fabrication techniques for Micro Robots. MEMS devices take very less volume and power with negligible mass while integrating both the mechanical and electronic devices. The techniques are very similar to those used in semiconductor device fabrications as the same accuracy is required in the microscale world as depicted below, shows some mems fabricated gears next to a spider mite. Some of the common techniques are deposition of material layer via methods like oxidation, implanting, diffusion, etc., and even combination of them).

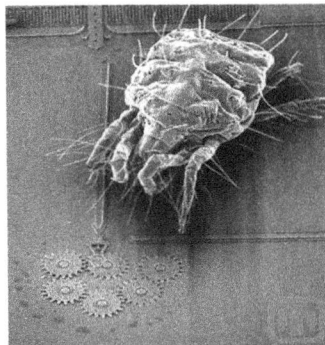

Size comparison of a MEMS gear & Spider Mite

Bulk microfabrication was the first MEMS fabrication technology. It employs fabrications on a single silicon crystal wafer, it is machined using etching and anodic bonding is done to work on the thickness of the silicon wafer. Unlike bulk, in surface microfabrication we selectively etch a silicon substrate to produce structures on top of the substrate. To overcome the issues of material handling at the micro scale, laser micro machining is used. It is used to precisely process piles of composite material both in cured and uncured state. Laser micromachining creates a localized heating spot for cutting through the substrate. Because of the use of epoxy, it thus becomes important to decide if cured or uncured substrates are used.

A very fast solution to fabrication and assembly is under research currently and that is printable robots. This technique would lead to inexpensive and rapid prototyping. Each printable robot would generally be in absolute 2 dimensional flat sheets as inspired from the Origami. Therefore, assembly is the other stage of printable robots. These robots are self-assembled by folding. In order to achieve local and sequential folding, the substrates at the bends have to be actuated by local stimulus. The basic principle used here is the Joules heating law. The cross-sectional area can we varied as required to the magnitude bend and heat dissipation controlled properly. Serpentine Trace patterns are used widely in foldable robots because their pattern help increase resistance and at the same time also distribute heat all over the surface area.

Fabrication of micro robots is best when done in their final form using bulk manufacturing or even self- assembly methods like folding, but if the components are fabricated separately it is also possible to assemble all these parts. Manual methods like tweezers can be used to assemble parts of about 100μm in size.

All the equipment being used at large scale mechatronic systems become more challenging to manufacturer as the size reduces, this is because the surface effects start dominating the Newtonian forces. This causes constraints on the performance and freedom of the bot depending on the materials available for fabrication and assembly at the MEMS scale.

The choice of materials affects the performance of the bot. Using high performance composite materials has provided a clear improvement in performance for micro robotics. Such materials overthrow common MEMS materials as they provide better fatigue properties and greater stiffness to weight ratios as compared to traditional materials. Hence increasing the overall performance.

Smart Composite Microstructures (SCM) technology is another fabrication technique which relies on the use of composite materials for the microfabrication . The desired compliance profile is generated of the required geometry and material properties by the help of laser micromachining on the constituent laminae. Laser micromachining is an effective technique for making thin sheets with micron-scale actuators and articulated laminates. The structure of the microrobot is created by sandwiching polymers between face sheets of rigid composite materials. These layers can we actuated whenever any of the laminae are being electro activated.

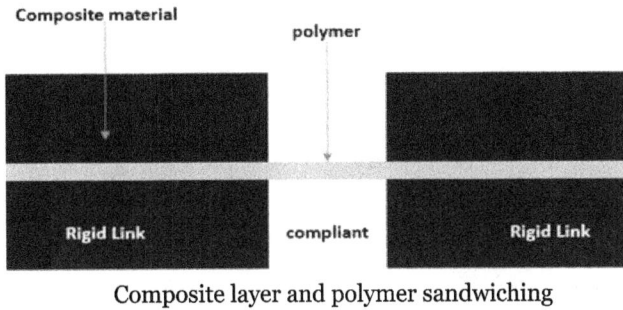

Composite layer and polymer sandwiching

Micro robots for biological applications raise a major concern over the biocompatibility of the micro robots. Most composite materials are not biocompatible. If the microrobot is being injected inside the human tissue, it should not cause any damage or infection to it.

Applications of Microrobotics

Cancer Fighting Robot

Research for developing micro robots that can have the ability to target cancer cells have been going on since long. Scientists are finally coming up with micro robots than can be used for fighting cancers. One such robot used to combat cancer is the Bacteriobot.

The bacteriobot is the world's 1st nanorobot for active medical treatment. These robots are genetically modified non-toxic bacteria attached to a bead that specifically attack tumour cells in the body. These robots are directly injected into the blood stream, diagnose and treat cancer by migrating and targeting tumours. The nano robot delivers the drug directly to the tumour and attacks the tumour leaving healthy cells alone. In this way it spares the patient from the side effects of chemotherapy.

The limitation of the bacteriobot is that it can only detect solid tumour forming cancers such as breast cancer and colorectal cancer.

Micro Robots in Eye Surgery

The eye is one of the most delicate organs in the human body. For years' research has been going on how to make eye surgeries more efficient and less time consuming.

One such discovery which redefined eye surgeries was the magnetically guided micro robot created at the Multi Scale Robotics Lab (MSRL) at ETH Zurich. Researchers at this lab named this micro robot as the OctoMag. The size of OctoMag can be imagined as the size of few human hairs i.e. 285μm. This robot is controlled externally and can carry delicate surgeries and removes the necessity of slicing the eye open for surgery.

The OctoMag is a magnetically manipulated system which consists of electromagnetic coils to wirelessly guide themicrorobots for eye surgery. Due to their small size theycannot carry batteries and motors so they are externally controlled using magnetic fields generated by OctoMag in 3D.

OctoMag

Nanorobotics

Nanorobotics describes the technology of producing machines or robots at the nanoscale. 'Nanobot' is an informal term to refer to engineered nano machines. Though currently hypothetical, nanorobots will advance many fields through the manipulation of nano-sized objects.

The field of medicine is expected to receive the largest improvement from this technology. This is because nanotechnology provides the advantage of transporting large amounts of nanorobots in a single injection. Furthermore, designs that include a communication interface will allow adaptations to the programming and function of nanobots already in the body. This will improve disease monitoring and treatment whilst reducing the need for invasive procedures.

Nanorobotic Applications in the Field of Hematology

Current research is developing nanorobotic applications for the field of hematology. This ranges from developing artificial methods of transporting oxygen in the body after major trauma to forming improved clotting capabilities in the event of a dangerous hemorrhage.

Respirocytes are hypothetical nanobots engineered to function as artificial red blood cells. In emergencies where a patient stops breathing and blood circulation ceases, respirocytes could be injected into the blood stream to transport respiratory gases until the patient is stabilized.

Current proposals suggest respirocytes would be able to supply 200 times more respiratory gas molecules than natural red blood cells of the same volume. Clottocytes are another type of nanobot which function as artificial platelets for halting bleeds.

Clottocytes would mimic the natural platelet ability to accumulate at the bleed, in order to form a barrier, by unfurling a fiber mesh which would trap blood cells when the nanobot arrives at the site of the injury. The clotting ability of one injection of clottocytes would be 10,000 times more effective that an equal volume of natural platelets.

Nanorobotics Applications for Cancer Detection and Therapy

As cancer survival rates improve with early detection, nanorobots designed with enhanced detection abilities will be able to increase the speed of a cancer diagnosis and therefore enhance the prognosis of the disease. Nanobots with embedded chemical sensors can be designed to detect tumor cells in the body. Proposed designs currently include the employment of integrated communication technology, where two-way signaling is produced. This means that nanobots will respond to acoustic signals and receive programming instructions via external sound waves along with transmitting data they have accumulated.

A simple reporting interface could be produced through strategically positioned nanobots in the body which are able to log information supplied by active nanobots traveling through the blood stream. Instructions could be adapted in vivo to provide active targeting for monitoring or healing.

Nanorobots with chemical sensors can also be utilized for therapy. Through specific programming to detect different levels of cancer biomarkers such as ecadherins and betacatenin, therapy can be provided in both primary and metastatic phases of cancer. Nanobots have the advantage of producing targeted treatment. Current cancer treatments have severe side effects caused by the destruction of healthy cells. Targeted treatment can be formed by designing nanorobots with chemotactic sensors on their surface which correspond to specific antigens on the cancer cells.

Nanorobotics Applications for Biohazard Defense

Nanorobots will also have useful applications for biohazard defense, including improving the response to epidemic disease. Nanobots with protein based biosensors will be able to transmit real-time information in areas where public infrastructure is limited and laboratory analysis is unavailable. This is particularly applicable for biomedical monitoring of areas devastated by epidemic disease as well as in remote or war torn countries during humanitarian missions.

Nanorobotics may also reduce contamination and provide successful screening for quarantine. In the event of an influenza epidemic for example, increased concentrations of alpha-NAGA enzyme in the blood stream could be used as a biomarker for the influenza infection. The increased concentration would trigger the nanorobot prognostic protocol which sends electromagnetic back propagated signals to portable technology such as a mobile phone. The information would then be retransmitted via the telecommunication system providing information on the location of the infected person, increasing the speed of contamination quarantine.

Swarm Robotics

Swarm robotics is the use of numerous, autonomous robotics to accomplish a task. Robot swarms coordinate the behaviors of a large number of relatively simple robots in a decentralized manner. Swarm robotics plays an important role in the development of collective artificial intelligence (AI). Current uses for robot swarms include search and rescue, precision agriculture, supply chain management (SCM) and military reconnaissance.

Swarm robotics attempts to draw on the ways social organisms, such as insects, use collaborative behaviors to achieve complex tasks beyond any individual's capability. For example, researchers in swarm robotics might study how bees mark trails with pheromones to map geographical locations. The researchers might then use the bee's algorithms to replicate that same behavior with robots.

Characteristics of Swarm Robotics

Since the swarm robotics is mostly inspired from the nature swarms, it's a good reference for analyzing the characteristics of swarm robotics through the characteristics of nature swarms. Counting from the first attempt on a swarm of intelligent agents, the research of swarm robotics started a century ago, taking the advantages of the nature swarm researches.

The first hypothesis of nature swarms is quite personified and assumes that each individual has a unique ID for cooperation and communication. The information exchange in the swarm is regarded as a centralized network. The queens in ant and bee colonies are supposed to be responsible for transmitting and assigning the information to each agent. However, S. Jha proved that the network in the swarm is decentralized. Thanks to the researches in recent half century, the biologists can now assert that there are no unique ID or other globally storage information in the network. No single agent can access to all the information in the network and a pacemaker is therefore inexistent.

The biologists now believe that the social swarms are organized as a decentralized system distributed in the whole environment which can be described through a probabilistic model. The agents in the swarm follow their own rules according to local information. The behaviors of the group emerge from these local rules which affect the information exchange and topology structure in the swarm. The rules are also the key component to keep the whole structure to be flexible and robust even when the sophisticated behaviors are emerged.

Application Scopes of Swarm Robotics

The potential applications of swarm robotics include the tasks that demand the miniaturization, like distributed sensing tasks in micro machinery or the human body. On the other hand, the swarm robotics can be suited to the tasks that demand the cheap designs, such as mining task or agricultural foraging task. The swarm robotics can be also involved in the tasks that require large space and time cost, and are dangerous to the human being or the robots themselves, such as post disaster relief, target searching, military applications, etc.

The study of robotics application in target search has grown substantially in the recent years. It is more preferable for the dangerous or inaccessible working area. The problems involved in swarm robotics research can be classified into two classes. One class of the problems is mainly based on the patterns, such as aggregation, cartography, migration, self-organizing grids, deployment of distributed agents and area coverage. Another class of problems focuses on the entities in the environment, e.g. searching for the targets, detecting the odor sources, locating the ore veins in wild field, foraging, rescuing the victims in disaster areas and etc. Besides these problems, the swarm robotics can also be involved into more sophisticated problems, mostly hybrid of these two classes, including cooperative transportation, demining, exploring a planet and navigating in large area.

Several potential application scopes for swarm robotics systems which are very suitable are described below, classified as mainly four types.

Tasks Cover Large Area

Swarm robotics system is distributed and specialized for the tasks requiring a large area of space, e.g. the tasks cover large areas. An easy example would be searching and collecting multiple targets in an open area. The swarm tries to search with group cooperation to accelerate the search. The area may be very large and the swarm can take advantage of the parallel searching with several small groups within the sensing ranges of the robots.

In another scenario, the robots in the swarm are distributed in the environment and can detect the dynamic change of the entire area, such as chemical leaks or pollution. The swarm robotics can complete such tasks in a better way than sensor network since each robot can patrol in an area rather than stay still. This means that the swarm can monitor the area with fewer individuals. Besides monitoring, the robots in the swarm can locate the source, move towards the area and take quick actions. In an urgent case, the robots can aggregate into a patch to block the source as a temporary solution. Another example.

Tasks Dangerous to Robot

Thanks to the scalability and stability, the swarm provides redundancy for dealing with dangerous tasks. The swarm can suffer loss of robots to a great extent before the job has to be terminated. The robots are very cheap and are preferred for the areas which probably damage the workers. In some tasks, the robots may be irretrievable after the task, and the use of complex and expensive robots are thus economically unacceptable while the swarm robotics with cheap individuals can provide the reasonable solutions. For example, Murphy et al. summarized the usage of robotics in mine rescue and recovery. They pointed out that although several applications already in use, the robots are beyond the requirement to show a desired performance in the tough environment under the ground. They proposed 33 requirements for the robots so as to achieve an acceptable behavior.

Tasks Require Scaling Population

Workload of some tasks may change over time, and the swarm size should be scaled based upon the current workload for high efficiency in both time and economics. For example, in the task of clearing oil leakage after tank accidents, the swarm should maintain a high population when the oil leaks fast at the beginning of the task and gradually reduce the robots when the leak source is plugged and the leaking area is almost cleared. The swarm also scales among different regions if the progress of these regions becomes unbalanced.

Tasks Require Redundancy

Robustness in the swarm robotics systems mainly benefits from the redundancy of the swarm, i.e. removing some robots does not have a significant impact on the performance. Some tasks focus on the result rather than the process, i.e. the system should make sure that the task will be completed successfully, mostly in the way of increasing redundancy.

References

- Robotics-32836: techopedia.com, Retrieved 11 July 2018

- Robotics-tutorial-introduction-robots: engineersgarage.com, Retrieved 29 March 2018

- Anthrobotics-where-the-human-ends-and-the-robot-begins: futurism.com, Retrieved 09 July 2018

- 7-bio-inspired-robots-mimic-nature, motion-control: machinedesign.com, Retrieved 09 July 2018

- Cognitive-robotics: ieee-ras.org, Retrieved 29 June 2018

- Four-types-of-robot-control-5131: brainkart.com, Retrieved 12 May 2018

- Developmental-robotics: scholarpedia.org, Retrieved 20 March 2018

- Swarm-robotics: searchenterpriseai.techtarget.com, Retrieved 11 April 2018

2

Common Types of Robots

Robots are used in diverse environments and for varied applications. Depending on the potential application, robots can be categorized as military robots, industrial robots, mobile robot, etc. This chapter has been carefully written to provide an overview of common robot types and their applications.

Autonomous Robots

Autonomous robots have the ability to gain information about their environments, and work for an extended period of time without human intervention. Examples of these robots range from autonomous helicopters to robot vacuum cleaners. These self-reliant robots can move themselves throughout the operation without human assistance, and are able to avoid situations that are harmful to themselves or people and property. Autonomous robots are also likely to adapt to changing surroundings.

Simpler autonomous robots use infrared or ultrasound sensors to see obstacles, allowing them to navigate around the obstacles without human control. More advanced robots use stereo vision to see their environments; cameras give them depth perception, and software allows them to locate and classify objects in real time.

Autonomous robots are helpful in busy environments, like a hospital. Instead of employees leaving their posts, an autonomous robot can deliver lab results and patient samples expeditiously. Without traditional guidance, these robots can navigate the hospital hallways, and can even find alternate routes when another is blocked. They will stop at pick-up points, and collect samples to bring to the lab.

DARPA, the Defense Advanced Research Projects Agency, is a division of the US Defense Department with the mission to create technological surprise for our enemies. This organization represents the cutting edge of military and disaster relief technology, and after developing autonomous vehicles, they are focused on creating autonomous robots that are capable of performing complex tasks in dangerous environments.

Another place autonomous robots are useful is in our natural environment. In 2013, researchers at Virginia Tech developed an autonomous robotic jellyfish with the intent of one day conducting undersea military surveillance or monitoring the environment. The 5 foot 7 inch jellyfish has a long duration and range of operation.

As emerging technologies become more prominent, the relationship between humans and robots is evolving. Autonomous robots have the ability to replace humans, such as a cognitive virtual assistant acting as an automated customer representative. Autonomous robots even have the ability to understand the emotion in a human's voice. These trends towards robotic involvement in industry processes will allow companies to improve productivity and customer experience, and gain a competitive advantage.

Types of Autonomous Robotic System

Programmable Automatic Robot

A programmable robot is a first generation robot with an actuator facility on each joint. The robots can be reprogrammable based on the kind of application they are commissioned to. The function and application of the robots can be changed by reprogramming after the robot is programmed once to perform a function in the given pattern and fixed sequence.

Robot kits like Lego mind storms, Bioloid from programmable Robotics can help the students to learn about its programming and working. The advanced mobile robot, robotic arms and gadgeteer are some of the examples of these programmable robots.

The main drawback of this autonomous robot is that once programmed it persists operation even if there is a need to change its task (in the event of emergency). These robots can be used in different applications like mobile robotics, industrial controlling and space craft applications.

Non-Programmable Automatic Robot

This robot is one of the basic types of robot, in fact, a non-programmable robot. This robot is not even considered as a robot, but is an exploiter lacking reprogrammable controlling device. The mechanical arms used in industries are some of the examples of these types of robots wherein the robots are generally attached to the programmable devices used in industries for mass production as shown in the figure.

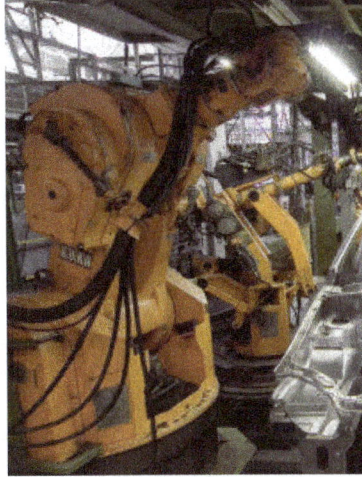

Non programmable automatic robot

These types of robots find applications in some of the devices including path guiders and medical products' carriers and also some line follower robots.

Adaptive Robot

Adaptive robots are also industrial robots that can be adapted independently to various ranges in the process. However, these robots are more sophisticated than programmable robots. These can be adapted up to a certain extent, and after evaluation they can perform the action required in that adapted area. These robots are mostly equipped with sensors and control systems.

Sensors are used to sense environmental conditions, process variables and other parameters related to a particular task. Feedback control system accesses these signals from the sensors, and depending on the algorithm implemented, it controls the outputs.

Adaptive robots

Adaptive robots are mainly used in applications such as spraying and welding systems. Robotic gripper and 2- finger adaptive gripper are examples of this autonomous robot. These robots can be used in different applications like aerospace, medical, consumer goods, house-hold applications and manufacturing industrial areas.

Intelligent Robots

Intelligent Robots

Intelligent robots, as the name suggests, are the most intelligent of all the other types of robots with sensors and microprocessors for storing and processing the data. These robots performance is highly efficient due to their situation-based analyzing and task performing abilities. Intelligent robots can sense the senses like pain, smell and taste and are also capable of vision and hearing, and – in accordance, perform the actions and expressions like emotions, thinking and learning.

These robots find their applications in the fields like medical, military applications and home appliance control systems, etc.

Aerobot

Aerobots are lighter-than-air robotic vehicles that could provide a platform for the exploration of planets and moons with an atmosphere, such as Venus, Mars, Titan and the gas giants. Aerobots have modest power requirements, extended mission duration and long traverse capabilities. They can execute regional surveys. The vehicles can transport and deploy scientific instruments and in-situ laboratory facilities over vast distances. They also can take wide-area surface samples.

There are three sorts of aerobots one type is an unmanned aerial vehicle (UAV), which is launched from an orbiting spacecraft and deploys wings before it lands. The second is a lighter-than-air craft, a helium-filled balloon fitted with heaters powered by solar cells. A third is a hybrid balloon-kite system sometimes called a helikite.

An advantage of the second and third kinds is that they don't need fuel, which is expensive to transport and which runs out all too soon. Unlike satellites or the winged aerobot, the lighter-than-air versions can fly for extended periods, allowing long-term monitoring of the properties of the atmosphere. They can also view and record the ground from much lower altitudes than is possible with satellites.

Android

Some androids are built with the same basic physical structure and kinetic capabilities as humans but are not intended to really resemble people. They may have jointed arms and legs, for example, that are capable of moving in the same ways that human limbs do, but have a plastic or metal exterior that in no way mimics human appearance. Examples of this type of android include Aldebaran Robotics' Nao and Google-owned Boston Dynamics' Atlas robot.

Other androids resemble humans so closely that they could be mistaken for living people; this type of android is often modeled on live humans. Eve-R, from the Korea Institute of Industrial Technology (KITECH) and Geminoid DK are two examples of this type of android.

Top: Atlas and Nao Bottom: Geminoid DK and EveR-1

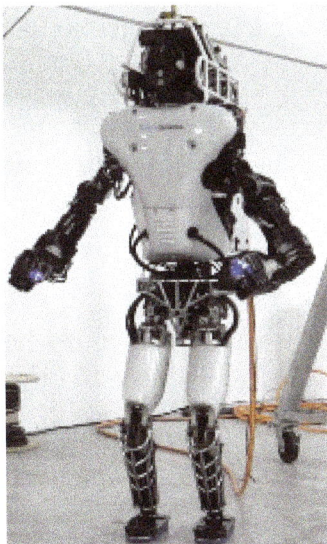

Some definitions differentiate between the two types of robots, referring to robots that only resemble people in basic form as humanoid robots and those that look like people as androids. According to other definitions, the terms are synonymous.

Examples of Android

Humanoid robots have come eerily close to overcoming the uncanny valley. With the right features in place, they are almost indistinguishable from their organic counterparts. Almost. The latest iterations are able to talk like us, walk like us, and express a wide range of emotions. Some of them are able to hold a conversation, others are able to remember the last interaction you had with them.

As a result of their highly advanced status, these life-like robots could prove useful in helping out the elderly, children, or any person who needs assistance with day-to-day tasks or interactions. For instance, there have been a number of studies exploring the effectiveness of humanoid robots supporting children with autism through play.

while the technology behind advanced android robotics has come a long way, there is still a lot of work to be done before we can have a face-to-face conversation with an entity without being able to tell that we are speaking with a replica.

But that is not to say that scientists and engineers haven't come close. With this in mind, here are six humanoid robots that have come the closest to overcoming the uncanny valley.

The First Android Newscaster

In 2014, Japanese scientists proudly unveiled what they claim to be the very first news-reading android. The life-like newscaster called "Kodomoroid" read a segment about an earthquake and an FBI raid on live television.

Although it – or she – has now retired to Tokyo's National Museum of Emerging Science and Innovation, she is still active. She helps visitors and collects data for future studies about the interactions between human androids and their real-life counterparts.

Yoshikazu Tsuno

BINA48

BINA48 is a sentient robot released in 2010 by the Terasem Movement under the supervision of entrepreneur and author Martine Rothblatt. With the help of robotics designer and researcher David Hanson, BINA48 was created in the image of Rothblatt's wife, Bina Aspen Rothblatt.

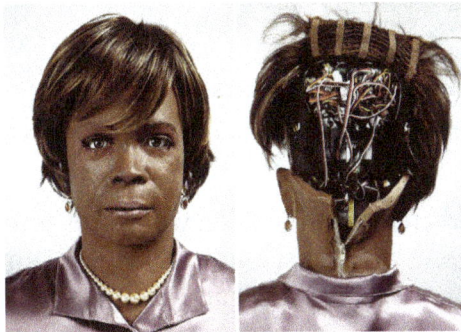

Hanson Robotics

BINA48 has done an interview with the New York Times, appeared in National Geographic and has traveled the world, appearing on a number of TV shows.

Geminoid DK

GeminoidDK

GeminoidDK is the ultra-realistic, humanoid robot that resulted from a collaboration between a private Japanese firm and Osaka University, under the supervision of Hiroshi Ishiguro, the director of the university's Intelligent Robotics Laboratory.

GeminoidDK is modeled after Danish professor Henrik Scharfe at Aalborg University in Denmark. Unsurprisingly, his work surrounds the philosophical study of knowledge – what separates true from false knowledge.

It is not only the overall appearance that was inspired by professor Scharfe. His behaviors, traits, and the way he shrugs his shoulders were also translated into life-life robotic movements.

Junko Chihira

Calenjapon

This ultra-realistic android created by Toshiba works full-time in a tourist information center in Tokyo. She can greet customers and inform visitors on current events. She can speak Japanese, Chinese, English, German, and even sign language.

Junko Chihira is part of a much larger effort by Japan to prepare for the 2020 Tokyo Olympics. Not only robotic tourist assistants will be helping the country with the incoming flood of visitors from across the globe in 2020; drones, autonomous construction site machines and other smart facilitators will be helping as well.

Nadine

NTUsg

This humanoid was created by the Nanyang Technological University in Singapore. Her name is Nadine, and she is happy to chat with you about pretty much anything you can think of. She is able to memorize the things you have talked to her about the next time you get to talk to her.

Nadine is a great example of a "social robot" – a humanoid that is capable of becoming a personal companion, whether it is for the elderly, children or those who require special assistance in the form of human contact.

Sophia

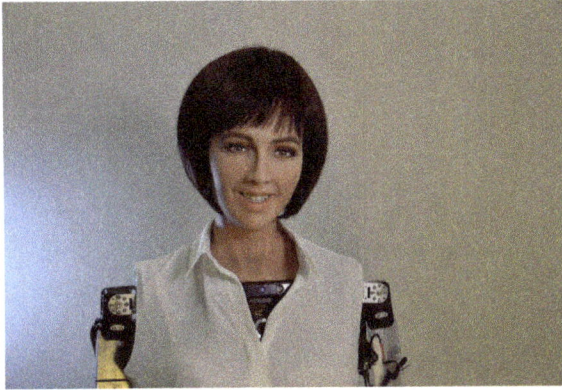

Hanson Robotics

Perhaps one of the most recent, most prominent life-like humanoids to be shown off in public is Sophia. You might recognize her from one of many thousands of public appearances, from The Tonight Show Starring Jimmy Fallon to SXSW. She was created by Hanson Robotics and represents the latest and greatest effort to overcome the uncanny valley.

She is capable of expressing an immense number of different emotions through her facial features and can gesture with full-sized arms and hands.

Ballbot

Ballbots are human-sized, dynamically stable mobile robots that balance on top of a single spherical wheel. They were originally developed to address fundamental problems with locomotion in statically stable mobile robots - statically stable mobile robots must have wide bases and low acceleration to avoid tipping over when moving. This results in slow, clunky, and fat robots. Dynamically stable mobile robots, such as ballbots, avoid this problem by balancing actively. Traditional drive mechanisms for ballbots have been mechanically complex, involving either omniwheels or timing belts and steel rollers.

Inverse Mouseball Drive

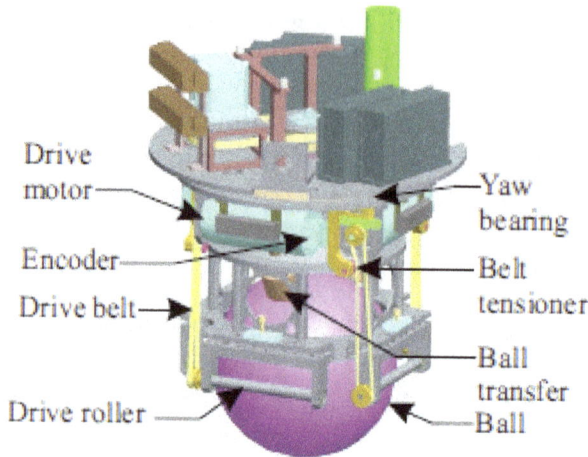

Inverse mouseball drive (IMB)

The CMU Ballbot relies on an inverse mouseball drive, or IMB. The IMB is shown in figure. Four timing belts connect dc motors to stainless steel rollers. The stainless steel rollers squeeze the polyurethane covered ball and transmit the motor rotation to the ball. This drive system has proven very successful over the years, but suffers from two main disadvantages. The first is that it is reasonably mechanically complex; for instance, the timing belts can wear and require maintenance. The second is that it has excessive friction. The rollers that are parallel to the direction of motion of the ball must slip along the ball's surface.

Omniwheel Drive Systems

The second primary type of drive system used in ballbots is an omniwheel-based system. In this drive system, typically three omniwheels support the weight of the robot and rest on top of the ball. The first use of omniwheels in a ballbot that the author is aware of is BallIP, shown in Figure. Rezero is a second example of an omniwheel-based system and was developed shortly after BallIP at ETH Zurich.

Omniwheel drive systems have their own set of advantages and disadvantages. They do not suffer from the same friction problem as the IMB and do not need timing belts. On the other hand, the omniwheels themselves are extremely mechanically complex and consist of hundreds of tiny moving parts. Furthermore, the omniwheels must be of a certain minimum size to have the load capacity to support the entire weight of the robot in designs with omniwheels on the top of the ball. This makes it difficult to obtain a sufficient speed reduction between the motors and the ball, which in turn often results in the use of gears. Note that the IMB benefits from a large gear reduction between the motors and the ball due to the presence of the stainless steel rollers. Gears introduce friction and backlash.

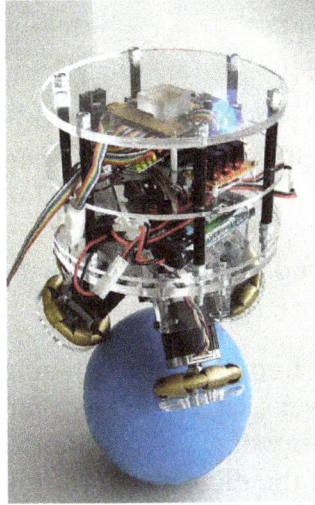

Additionally, the body of the robot can turn freely with respect to the ball; this introduces an additional control challenge as the robot must use its drive system to maintain a constant heading. In the IMB, the body does not turn freely with respect to the ball and a separate yaw mechanism is used to control the heading of the robot.

Spherical Induction Motor

The spherical induction motor (SIM) in figure below. It consists of a hollow steel ball with an outer copper shell which rests on a set of six passive nylon ball transfers inside the motor frame. Six sta- tors are positioned around the circumfer- ence of the ball. Each stator is driven by a custom three-phase motor driver which controls the amount of force applied to the ball by the stator.

The SIM is extremely mechanically simple and the motivation for using it as the drive system for a ballbot is to create a high- performance mobile robot with just two moving parts.

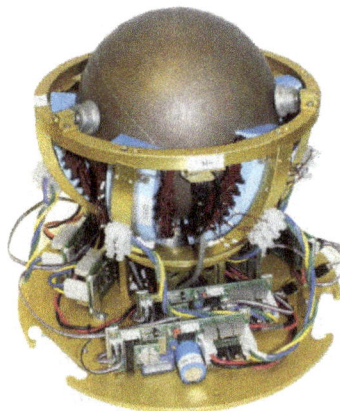

Spherical induction motor (SIM)

Hexapod

Legged hexapod robots are programmable robots with six legs attached to the robot body. The legs are controlled with a degree of autonomy so that the robot can move within its environments, to perform intended tasks. Hexapod robots can be suitable for terrestrial and space applications, and they can include features such as omnidirectional motion, variable geometry, good stability, access to diverse terrain, and fault tolerant locomotion.

One of the motivating factors often given for pursuing the development of hexapod robots is that they can climb over obstacles larger than the equivalent sized wheeled or trucked vehicle. In fact, the use of wheels or crawlers limits the size of the obstacle that can be climbed to half the diameter of the wheels. On the contrary, legged robots can overcome obstacles that are comparable with the size of the machine leg. Hexapod walking robots also benefit from a lower impact on the terrain and have greater mobility in natural surroundings. This is especially important in dangerous environments like mine fields, or where it is essential to keep the terrain largely undisturbed for scientific reasons. Hexapod legged robots have been used in exploration of remote locations and hostile environments such as seabed, in space or on planets in nuclear power stations, and in search and rescue operations. Beyond this type of application, hexapod walking vehicles can also be used in a wide variety of tasks such as forests harvesting, in aid to humans in the transport of cargo, as service robots and entertainment.

Despite the above referenced aspects, many challenges remain before hexapod walking robots can have a more widespread use. Some of their current disadvantages include higher complexity and cost, low energy efficiency , and relatively low speed. Walking robots are in fact complex and expensive machines, consisting of many actuators, sensors, transmissions and supporting hardware.

Recent Developments

The two last decades have been characterized by a rapid development of control systems technology. Hexapod robots were equipped with various sensing systems. Artificial Intelligence systems were widely applied to the analysis of environment and motion of robots on a complex surface.

A series of bio inspired robots was developed at Case Western Reserve University (USA) at the end the 90s, such as, for example, Robot III that had a total of 24 DoFs. Robot III architecture was based on the structure of cockroach, trying to imitate their behavior. In particular, each rear leg had three DoFs, each middle leg four DoFs and each front leg five DoFs. Similarly, Biobot was a biomimetic robot physically modeled as the American cockroach (Periplaneta Americana) and powered by pressurized air.

This hexapod had a great speed and agility. Each leg of the robot had three segments, corresponding to the three main segments of insect legs: coxa, femur, and tibia.

Hamlet was a hexapod robot constructed at the University of Canterbury, New Zealand. Its legs were all identical and each had three revolute joints. Hamlet's application task was to study force and position control on uneven terrain.

In 2001, a project named RHex commenced RHex design comes from a multidisciplinary and multi-university DARPA funded effort that applies mathematical techniques from dynamical systems theory to problems of animal locomotion. Hexapod design consists of a rigid body with six compliant legs, each with one DoF. Thus, RHex has only six motors that rotate the legs such as a wheel.

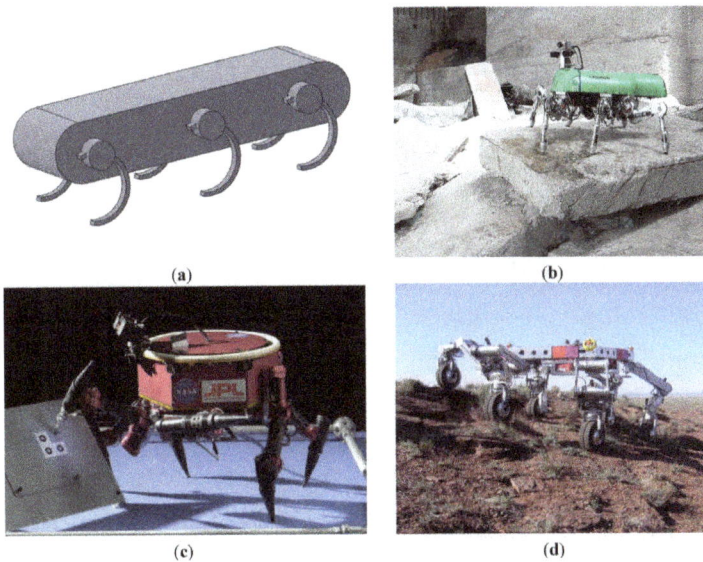

Recent developments in hexapod design: (a) RHex basic architecture; (b) Six-legged walking robot LAURON V from the FZI Research Center for Information Technology in Karlsruhe, Germany; (c) One configuration of the LEMUR prototype: the front limbs function as legs and arms, with integrated tools; (d) ATHLETE during experimental tests.

Several prototypes of Rhex have been developed. At present the project is still active.

Lauron V hexapod robot was the result of about 10 years of progressive improvement on the previous configurations Lauron I, II, III and IV. LAURON is biologically-inspired by the stick insect. Like this insect, the robot has six legs fixed to a central body. Each of the six legs is actuated by four joints. Each foot has a three-axis force sensor, and each motor has a current sensor that detects forces opposing to its movement. At present the project is still active.

Gregor I reproduces the cockroach's agility where the locomotion control is based on the theory of the Central Pattern Generator. Gregor I had a biological inspiration where each leg pair has a unique design. The front leg pair and the middle leg pair have three DoFs on each leg, and the rear leg pair has two DoFs. Another hexapod robot called Sprawlita was developed following the basic principles of locomotion of cockroaches: self-stabilizing posture, different functions for the legs, passive visco-elastic structure, open-loop control and integrated construction. In 2005, the hexapod robot named BILL-Ant-p was developed. The robot was based on ants' behavior and it is composed of three DoFs on each leg with six force-sensing feet, a three-DoF neck and head, and actuated mandibles with force-sensing for a total of 28 DoFs.

A series of hexapod named LEMUR (Limbed Excursion Mechanical Utility Robots robot) was developed by Jet Propulsion Laboratory with the goals of using robots for repair and maintenance in near-zero gravity on the surface of spacecraft. MARS (Multi Appendage Robotic System) was a hexapod mobile robotic research platform developed after the LEMUR project for similar applications by employing radial symmetry. MARS platforms were capable of walking in any direction without turning.

In 2004, a six-legged lunar robot called ATHLETE was developed by the Jet Propulsion Laboratory. This robot had the ability to roll rapidly on rotating wheels over flat smooth terrain and walk carefully on fixed wheels over irregular and steep terrain. ATHLETE had a payload capacity of 450 kg, a diameter of around 4 m and a reach of around 6 m.

AQUA was an amphibious hexapod robot developed with six independently-controlled leg actuators . One of the most important features of this robot was the ability to switch from walking to swimming gaits as it is moving from a sand beach or surf-zone to deep water. The underwater walking robot CR200 was built as based on the concept of Crabster. The application field was inspection of shipwrecks or scour and survey of seafloor in high current and turbid environments. The hexapod robot called RiSE is able to climb on a variety of vertical surfaces as well as demonstrating horizontal mobility. A mechanism applied in Rise uses compliant micro spines on its feet to reliably attach to textured vertical surfaces, in order to carry the payload while the robot climbs. COMET hexapods are a serious of robots designed to operate on extremely unstructured terrain. The latest prototype, COMET IV, is a hydraulically driven hexapod involving walking with the force/impedance control, fully autonomous navigation with laser mapping and teleoperation. Mantis is a hexapod robot hydraulic powered developed by Micromagic Systems. It stands nearly 3 m tall and weighs about 2 tons; at present, it is one of the biggest hexapod robots in the world.

Many hexapod rovers for space exploration have been designed and built in the last years. A very rich reference of this vehicle and their design issues can be found in.

Given the wide variety of existing hexapod robots, several of the above described prototypes have been detailed in a specific table. It provides a quick comparison tool for referring to the main features.

| Table: Comparison of hexapod robots. | | | | | | | | | | |
| Main Characteristics | | | | | | Main Performance | | | | |
Robot Name	Mass (Kg)	Length (cm)	Width (cm)	Height (cm)	Total DoFs	Max Speed (m/s)	Gait/Mobility	Power (W)	Year	Application Tasks
Ambler	2700	500	500	700	12	0.007	Wave Free	1900	1989	Planetary exploration
ASV	3200	520	240	300	15	1.0	Wave Free	26000	1989	Navigate on uneven terrains
Hannibal	2.7	35	NA	20	19	0.04	Wave Free	NA	1989	Planetary exploration
Tum	23	80	40	100	18	0.3	Wave	500	1991	Hexapod following biological principles
Biobot	11	58	14	23	18	NA	NA	NA	2000	Locomotion over rough terrain
Hamlet	13	40	28	40	18	0.1	Wave Free	52	2001	Testing force and position control
Rhex	7	53	20	15	6	0.55	Wave Free	100	2001	Hexapod with reduced actuators
Sprawlita	0.27	16	NA	NA	12	0.35	Wave	NA	2002	Robots inspired to cockroaches
Lauron III	18	50	30	80	18	0.4	Wave Free	NA	1999–2003	Testing a hierarchical walk controller
Genghis	1	40	15	NA	12	0.04 0.12	Wave Free	NA	2004	Developing of a reactive controller
Aqua II	16.5	64	44	13	6	0.71	Land Water	200	2010	Underwater hexapod robot
Bill-Ant-p	2.3	47	33	16	22	0.004	Wave	25	2005	Biologically inspired legged robot
Gregor I	1.2	30	9	4	16	0.03	NA	25	2006	Robot inspired on cockroaches.
Athelete	850	2.75	2.75	2	36	2.78				
0.016	Wheleed									

Wave	NA	2006	Navigate on rough soil of on the Moon							
RiSe	2.8	41	NA	NA	12	NA	Wave	NA	2006	Hexapod climbing robots
Comet IV	2120	280	330	250	24	0.278	Wave	20600	2009–2011	Hexapod for multitasks on outdoor environment
CR200	600	250	200	130	18	0.5	Wave	20000	2013	Walk either on land or underwater in the turbulent surf zone
Mantis	1900	420	220	280	18	NA	Wave	42000	2013	Entertainment
NA means data not available.										

Hexapod Robots' Performance Indices

The following indices have been proposed in the literature in order to compare legged robots which have different masses, shape and sizes:

- Duty factor
- Froude number
- Specific resistance
- Stability margin
- The duty factor β is defined as:

$$\beta = \frac{\text{support period}}{\text{cycle time}}$$

Duty factor can be used to make the distinction between walks and runs, since we have $\beta \geq 0.5$ for walking and $\beta < 0.5$ for running.

According to , a Froude number can be calculated by:

$$F_{r2} = \frac{V^2}{gh}$$

where V is the walking or running speed, g is the acceleration due to gravity, and h is the height of hip joint from the ground. V is a characteristic speed of the motion. According to, a possible way to estimate V can be by using the products (h times f) instead of the characteristic speed V since h is the characteristic height of the leg and f is the stride frequency. Thus, one can write for hexapod robots:

$$\frac{\left(hf\right)^2}{gh} = \frac{hf^2}{g} = F_{r2}$$

Alexander used the Froude number in order to characterize animal locomotion; he showed that animals of different sizes use similar gaits when they travel with equal Froude numbers. In particular, most animals change their gait from walking to running at a speed equivalent to a Froude number of F = 1.

The specific resistance is a dimensionless number that is used to evaluate the energy efficiency of a mobile robot. Gabrielli and von Karman discussed the performance of various vehicles using the power consumption per unit distance.

That is:

$$\varepsilon = \frac{E}{M\,gd}$$

where E is the total energy consumption for a travel of distance d, M is the total mass of the vehicle, and g is the acceleration due to gravity. We can say that the specific resistance indicates how smooth the locomotion is.

For a dynamic walking robot, we can define a stability margin as the minimum distance of the ZMP to the boundaries of the support polygon, since the ZMP is the natural extension of a projected CoM on the ground. For a legged robot on rough terrain, Messuri and Klein defined the energy stability margin as the minimum potential energy required to tumble the robot.

Design Considerations

Designing hexapod legged robots is far from trivial. A very numerous and a wide range of possibilities exist to design a hexapod. Designers must take several decisions which influence the operation and technical features. Some of the most important design issues and constraints according to can be outlined as:

- The mechanical structure of robot body;

- Leg architecture;

- Max sizes;

- Actuators and drive mechanisms;

- Control architecture;

- Power supply;

- Walking gaits and speed;

- Obstacle avoidance capability;

- Payload;

- Autonomy;

- Operation features;

- Cost.

Industrial Robot

An industrial robot is an automatically controlled, reprogrammable, multipurpose manipulator programmable in three or more axes.

Main Parts of an Industrial Robot

An industrial robot arm includes these main parts: Controller, Arm, End Effector, Drive, Sensor

The controller is the "brain" of the industrial robotic arm and allows the parts of the robot to operate together. It works as a computer and allows the robot to also be connected to other systems. The robotic arm controller runs a set of instructions written in code called a program. The program is inputted with a teach pendant. Many of today's industrial robot arms use an interface that resembles or is built on the Windows operating system.

Industrial robot arms can vary in size and shape. The industrial robot arm is the part that positions the end effector. With the robot arm, the shoulder, elbow, and wrist move and twist to position the end effector in the exact right spot. Each of these joints gives the robot another degree of freedom. A simple robot with three degrees of freedom can move in three ways: up & down, left & right, and forward & backward. Many industrial robots in factories today are six axis robots.

The end effector connects to the robot's arm and functions as a hand. This part comes in direct contact with the material the robot is manipulating. Some variations of an effector are a gripper, a vacuum pump, magnets, and welding torches. Some robots are capable of changing end effectors and can be programmed for different sets of tasks.

The drive is the engine or motor that moves the links into their designated positions. The links are the sections between the joints. Industrial robot arms generally use one of the following types of drives: hydraulic, electric, or pneumatic. Hydraulic drive systems give a robot great speed and strength. An electric system provides a robot with less speed and strength. Pneumatic drive systems are used for smaller robots that have fewer axes of movement. Drives should be periodically inspected for wear and replaced if necessary.

Sensors allow the industrial robotic arm to receive feedback about its environment. They can give the robot a limited sense of sight and sound. The sensor collects information and sends it electronically to the robot controlled. One use of these sensors is to keep two robots that work closely together from bumping into each other. Sensors can also assist end effectors by adjusting for part variances. Vision sensors allow a pick and place robot to differentiate between items to choose and items to ignore.

Applications of Industrial Robot

Industrial robots are used in a variety of applications. These include:

- Handling: Capable of manipulating products as diverse as car doors to eggs, industrial robots are fast and powerful as well as dexterous and sensitive. Applications include pick and place from conveyor line to packaging, and machine tending, where raw materials are fed by the robot into processing equipment such as with injection molding machines, CNC mills and lathes and presses.

- Palletizing: Industrial robots load corrugated cartons or other packaged items onto a pallet in a defined pattern. Robotic palletizers rely on a fixed position or overhead gantry robot with special tooling that interfaces with the individual load components, building simple to complex layer patterns on top of a pallet that maximize the load's stability during transport. There are three primary types of palletizing: inline or layer forming, depalletizing or unloading, and mixed case.

- Cutting: Due to their dangerous nature, laser, plasma and water jet cutters are frequently used with robots. Hundreds of different cutting paths can be programmed into the robot, which produces precise accuracy and path following with greater flexibility than most dedicated cutting machines.

- Finishing: Multi-axis robots can grind, trim, fettle, polish and clean almost any part made in any material for a consistent quality finish.

- Sealing and gluing: To apply sealant or glue, a robot follows a path accurately with good control over speed while maintaining a consistent bead of the adhesive substrate. Robots are frequently used for sealing applications in the car industry to seal in windows, as well as in packaging processes for automated sealing of corrugated cases of product.

- Spraying: Due to the volatile and hazardous nature of solvent-based paints and coatings, robots are used in spray applications to minimize human contact. Paint robots typically have thin arms because they don't carry much weight, but need maximum access and movement fluidity to mimic a human's application technique.

- Welding: Used for both seam (MIG, TIG, arc and laser) and spot welding, robots produce precise welds, as well as control parameters such as power, wire feed and gas flow.

Uses of Industrial Robots

Robots are used in a variety of ways throughout manufacturing and distribution:

- Load building: Assembling a pallet load of products at the end of a production line.

- Manufacturing: Performing processing and assembly functions to work-in-process.

- Quality control: Testing and inspection procedures deploy robots for repetitive or dangerous work.

- Transportation: Loading pallets prior to shipping.

- Warehousing: Removing received products from pallets and routing them to storage locations within a facility.

Benefits of Industrial Robots

Industrial robots provide a variety of benefits:

- Accuracy – Robotic palletizers are software-directed for proper load placement.

- Flexibility – Robotic systems can be re-purposed for other uses; end effectors can be switched out to handle different load types.

- Lower labor costs – Automated pallet building reduces worker strain and frees operators for other tasksv.

- Quiet operation – Servo-based, robotic palletizers generate low noise levels.

- Reduced product damage – Gentle handling prevents package and product damage.

- Speed – The systems increase rate productivity up to 50%.

Industries where Industrial Robots are Used

Industrial robots are used in many industries, including:

- Aerospace
- Automotive
- Beverage
- Computers
- Consumer goods
- E-Commerce
- Electronics
- Food
- Grocery
- Hardware
- Healthcare
- Liquor distribution
- Manufacturing
- Medical products
- Pharmaceutical
- Quality control and inspection
- Retail
- Warehousing and distribution

Microbot

A microrobot is a miniaturized, sophisticated machine designed to perform a specific task or tasks repeatedly and with precision. Microrobots typically have dimensions ranging from a fraction of a millimeter up to several millimeters.

A microrobot, like its larger and smaller cousins, the robot and the nanorobot , can be either autonomous or insect-like. An autonomous microrobot contains its own on-board computer, which controls the machine and allows it to operate independently. The insect scheme is more common for microrobots. In an insect-microrobot arrangement, the machine is one of a fleet of several, or many, identical units that are all controlled by a single, central computer. (The term insect comes from the fact that such robots behave like ants in an anthill or bees in a hive.

Military Robot

The utilization of Robotics in military is well shown by US army. Osama and other terrorists were tracked by these military robots. They are robust, they are obedient, they are daring, they don't have fear of death, and most important they have proved themselves in Iraq and Afghanistan. Now, terrorists are terrified by drone attacks. The utilization of robotics technology in military led to a new field in robotics i.e. Military Robotics.

Representational Image of a Military Drone

Military robotics isn't about creating an army of humanoids but utilization of robotics technology for fighting terror and defending the nation. Thus, military robots need not be humanoids or they not necessarily need to carry weapons, they are just those robots that can help the armed forces. The opportunities offered by these technologies are boundless. Apart from army research centers there are many private firms also which provide military robots for defense forces like Foster Miller, 21st Century Robotics, EOD Performance, Northrop Grumman, and General Atomics etc. They have created many job opportunities and are developing this sector.

Operations

Today military robots use very sophisticated and advance technology for operations.

They use different technologies for reconnaissance, guidance and weaponry. They basically use GPS, Fiber Optic Tethers, LIDARs for guidance. GPS is based on satellite connections and is even used in mobile phones. The fiber optics is a hi-tech and hi-speed communication system especially used by defense. LIDARs are based on laser communication and nowadays used by traffic police to detect over speeding vehicles. For reconnaissance they use other technologies like cameras, electronic RF sensor, RA-DAR, etc. The robots are mainly used for reconnaissance purposes but they can also carry lethal and non-lethal weapons like AGM-114 Hellfire missiles, M249 saw machine guns, ammo can, bomb diffusal kits, grenades, etc.

Soviet TT-26 Teletank, 1940

Varieties

Military robots come in different shapes and sizes as per the task they are designated for. In the development of military robots, we can consider US Mechatronics which has created or developed a working automated sentry gun and is presently developing it further for commercial as well as military use. As far as military robots development is concerned, we cannot forget MIDARS which is a four-wheeled military robot. This robot is outfitted with many cameras, radar, and a firearm that performs arbitrary patrols around a military base automatically. Their size can vary from a small bot TALON and large UAV MQ-1 Predator. Their design is also task specific like, predator is for surveillance and attack from air so it is more like an airplane while TALON is for attack from ground so it is more like an armoured tank. There are three popular classes of military robots i.e. UGVs, UUVs and UAVs.

A Typical Military Robot For Ground Attack

UGVs

UGVs i.e. Unmanned Ground Vehicle are those which attack from ground. They have various sensors, cameras, arms mounted on them. The UGV like big-dog has been devised as four legged bot it can carry heavy loads that were used to be carried by soldiers. It has capability to carry load on uneven terrain. Robots like packbots are so compact, light and robust that army can carry them on their back. Controlled by a Pentium processor that has been designed specially to withstand rough treatment, Packbot's chassis has a GPS system, an electronic compass and temperature sensors built in. Packbot manufacturer iRobot says Packbot can move more than 8 mph (13 kph), can be deployed in minutes and can withstand a 6-foot (1.8-meter) drop on concrete -- the equivalent of 400 g's of force. Its design alows it to flip also.

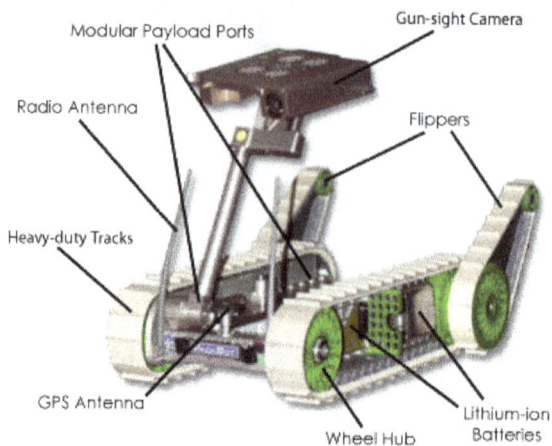

An Image Showing UGVs Packbot Robot With Diffrent Parts

The other most popular robot is TALON. It is manufactured by Foster Miller. The TALON is a man-portable robot operating on small treads. It weighs less than 100 lbs (45 kg) in its base configuration. TALON is operated with a joystick control, has seven speed settings (top speed is 6 feet/1.8 meters per second) and can use its treads to climb stairs, maneuver through rubble and even take on snow.

An Image Of TALON Robot Belong To UGV's

Versatility has been designed into the TALON as well, with multiple possible configurations available that adapt the robot to the situation at hand. The basic TALON includes audio and video listening devices and a mechanical arm. A lightweight (60-lb/27-kg) version omits the arm. TALONs were used for search and rescue at WTC Ground Zero, and they have been used in Bosnia, Afghanistan and Iraq for the disposal of live grenades, improvised explosive devices and other dangerous explosives.

UAVs

The UAVs are those which are used from air. They shouldn't be confused with missiles. Actually missiles are weapons but UAVs aren't weapons itself but they carry weapon. Some of the UAVs are RQ-11A/B Raven, RQ-5A / MQ-5A/B Hunter, MQ-1 Predator , MQ-9 Reaper etc. Northrop Grummanand General Atomics are the dominant manufacturers in UAV industry. There are other countries except USA also in UAV technology like Israel, Europe, India, etc. which have successfully developed UAVs and are developing it further.

The military uses several different flying robots, mainly for reconnaissance. Instead of UGVs, these are known as UAVs (unmanned aerial vehicles), and they are sometimes referred to as drones. UAVs look like model aircraft, and they range in size from small planes that can be held by a person and launched with a good throw, like the FQM-151 Pointer, to full-size airplanes that operate by remote control, like the RQ-4A Global Hawk.

RQ-4A Global Hawk

Global hawk specifications:

An image Of RQ-4A Global Hawk In UAV's Military Robot

- Length: 44 ft. 4.75 in (13.53 m)
- Wingspan: 116 ft. 2.5 in (35.42 m)
- Height: 15 ft. 2.5 in (4.64 m)
- Weight empty: 14,800 lb. (6,710 kg)
- Weight max: 25,600 lb. (11,600 kg)
- Speed: 403 mph (648 kph)
- Ceiling: 65,000 ft. (19,800 m)
- Range: 11,730 nautical miles (21,720 km)
- Endurance: 36 hours
- Propulsion: Rolls-Royce/Allison F137-AD-100 turbofan

FQM-151 Pointer

Pointer Specifications

- Length: 6 ft. (1.83 m)
- Wingspan: 9 ft. (2.74 m)
- Weight: 9.6 lb. (4.3 kg)
- Speed: 50 mph (80 kph)
- Ceiling: 985 ft. (300 m)
- Mission radius: 2.7 nautical miles (5 km)

- Endurance: Primary batteries - 1 hour; Rechargeable batteries - 20 min

- Propulsion: Electric motor

A still Of FQM-151 Pointer In UAV's Military Robot

UUVs

The next class is UUVs i.e. unmanned underwater vehicle. These marvels have capabilities to be operated underwater. UUVs were envisioned to contribute to the following SSN mission areas: Mine Warfare (MIW); Intelligence, Surveillance, and Reconnaissance (ISR); and undersea environmental sensing and mapping. The Talisman UUV is a fully autonomous unmanned mini-sub designed and made by BAE Systems. It has the capability to perform shoreline reconnaissance when the risks may be deemed too high, or conditions unsuitable, for human operators. The Talisman system comprises a vehicle and remote control console. The vehicle can be re-programmed in mid-mission by satellite and features a carbon fibre hull giving it a stealth-like profile and low observability. Talisman can carry a wide variety of payloads, such as image capture, environmental sensors or mine counter measures.

An Image of Talisman UUV Designed By BAE Systems

Predator

Predator Drone UUV Military Robot

Specifications	
Manufacturer	General Atomics
Runway (ISA)	Improved, 3000 ft x 100 ft
RATO	No
Dimensions	length: 27 ft wingspan: 48.7 ft height: 6.9 ft weight: 1,130 lbs (empty)
A/Vs in Baseline	4
Data Link	Frequency: C-band Line of Sight: yes Satellite: yes (UHF and Ku) GPS: yes
Range	Line of Sight on GCS Unlimited via satellite
Power Plant	Rotax 912 pusher
Fuel	Type: 110 LL avgas capacity: 110 liters
Endurance	24 hours on station at 500NM 40 hours
Payloads	electro-optical IR (day-night) SAR(all weather) modular
Operational Altitudes (ISA)	Up to 26,000 ft MSL
Speeds	Stall: 54 kts Cruise: 70-90 kts Dash: 120 kts

Max Gross Take-off Weight	2250 lbs
Weather limits	T/O cross winds: 15 kts T/O head winds: 30 kts rain: No Visible moisture: limited Icing: No Turbulence: Light
Automatic Return Home on Data Link Loss	Yes
Emergency Recovery	Parachute
Relay Flight Capability	In development
Multiple AV Control	Yes
Multiple AV Control	Yes, autonomous flight
Control Transfer to Another Shelter	Yes
Role	Remote controlled, UAV
Manufacturer	General Atomics Aeronautical Systems
First flight	January 1994
Introduction	July 1995
Status	Active
Primary user	United States Air Force
Produced	1995–Present
Number built	360 (285 RQ-1, 75 MQ-1)
Unit cost	~ $4.5 million
Developed from	General Atomics GNAT
Variants	MQ-1C Grey Eagle
Developed into	MQ-9 Reaper

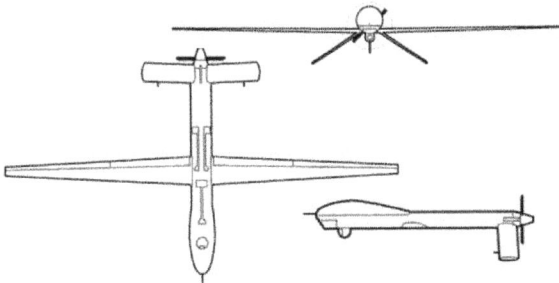

Diagram Explaining Structure And Design Of Predator Drone In Military

Structure and Design of Predator

The Predator uses a lightweight, 4-cylinder snowmobile engine, which powers a rear-mounted propeller, making the Predator a "pusher"-type aircraft. It carries cameras and other sensors but has been modified and upgraded to carry and fire two AGM-114 Hellfire missiles or other weapons. Hence, it's no wonder that technology is moving forward in the direction of creating army of robots that will not just make our life comfortable but also safe.

Expectations

In the near future robotics will reach new heights. All the dreams where robots do our jobs will come true. And military is not an exception. At present there are many hostile and painful tasks that have to be performed by humans but in near future most of them will be done by them. In future news headlines would be like"5 robos saved bank robbery", "a robo team has vanished terror camp, no casualties reported" and many more. This future is very close.

Now a variety of military robots is under consideration. TAC i.e. Tactical Autonomous Combatant is capable in working on ground, air, space, or undersea environments, and in humiliating conditions of extreme heat or cold. Further they will be capable of operating in chemically, biologically, or radioactive contaminated environments. As the name suggests TAC be autonomously operated or with very little human intervention though humans have to guide them but humans have to just guide them from miles away.

Mobile Robot

Mobile robotics is the industry related to creating mobile robots, which are robots that can move around in a physical environment. Mobile robots are generally controlled by software and use sensors and other gear to identify their surroundings. Mobile robots combine the progress in artificial intelligence with physical robotics, which allows them to navigate their surroundings.

Two types of mobile robotics are autonomous and non-autonomous mobile robots. Autonomous mobile robots can explore their environment without external guidance, while guided robots use some type of guidance system to move. Other semi-stationary robots have a small range of movement.

The field of mobile robotics is important today as companies look at how to utilize artificial intelligence. For instance, those looking at the consumer electronics world might remark that although autonomous mobile robots have been around for a while, they are not often represented in consumer offerings – for instance, in items like Alexa and other virtual tools that are stationary rather than robotic. One could ask when the new AI robotic consumer products are going to make their debut in global markets.

Nanobot

A nanorobot is a tiny machine designed to perform a specific task or tasks repeatedly and with precision at nanoscale dimensions, that is, dimensions of a few nanometers (nm) or less, where 1 nm = 10-9 meter. Nanorobots have potential applications in the assembly and maintenance of sophisticated systems. Nanorobots might function at the atomic or molecular level to build devices, machines, or circuits, a process known as molecular manufacturing. Nanorobots might also produce copies of themselves to re-place worn-out units, a process called self-replication.

The Mechanics of Nanobots

These minute molecules have components that enable them to identify and attach themselves to a cancer cell.

When activated by light, the nanobots' rota-like chain of atoms begin to spin at an incredible rate – around two to three million times per second. This causes the nanobot to drill into the cancer cell, blasting it open.

The study is still in its early stages, but researchers are optimistic it has the potential to lead to new types of cancer treatment.

Dr Robert Pal, of Durham University, said: "Once developed, this approach could provide a potential step change in noninvasive cancer treatment and greatly improve survival rates and patient welfare globally".

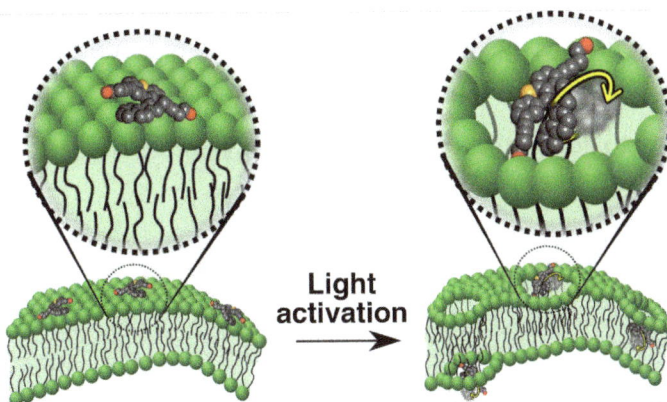

Light activation

The spinning nanobots burrow into cancer cells to destroy them

Nanobots in Our Veins

The destructive properties of the nanobots make them perfect for killing cancer cells. But the technology can also be used to repair damaged or diseased tissues at a molecular level.

In the future, these nanomachines could essentially patrol the circulatory system of the human body. They could be used to detect specific chemicals or toxins and give early warnings of organ failure or tissue rejection.

Another potential function may involve taking biometric measurements to monitor a person's general health.

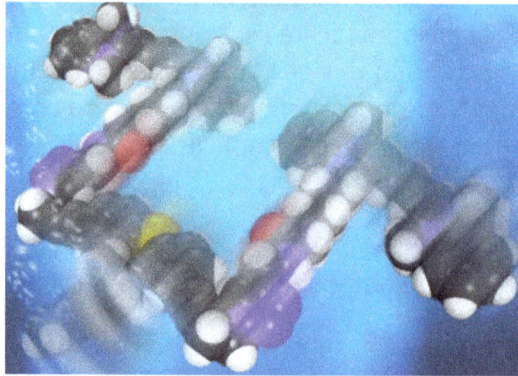

A computer-generated image of a nanobot

Searching for Oil

The medicinal advantages of nanobots are clear to see, but industry might also benefit from the technology.

Oil and gas is one example. The idea is that nanobots could be injected into geologic formations thousands of feet into the earth. Changes to the chemical make-up of the machines would point to the location of reservoirs.

Meanwhile, it's also been suggested that nanotechnology could become a valuable tool in cleaning up oil spills.

There is a long way to go before this tiny technology enters the mainstream, but it has the potential to make a massive impact on the world.

Rover

In 2004, NASA engineers and scientists successfully landed two robotic rovers, Spirit and Opportunity, on Mars to explore the terrain and gather evidence to determine the planet's history. Working as a team, engineers from several disciplines collaborated to achieve the complex task of designing and fabricating these two rovers. Electrical engineers designed and built the compact rover circuitry, mechanical engineers designed the body and arm, and computer engineers developed the rover programming and communication.

Spirit and Opportunity have many parts that are analogous to the human body. According to NASA, each Mars rover has a body, brains, a neck, a head, eyes, an arm, and wheels and legs, among other various parts and components. The body is used to protect their "vital organs" (also known as electronics) from the extreme temperatures of Mars. From day to night, temperatures can vary from 113 °C (235 °F) to -96 °C (-140 °F). The rover's brains are located inside of its body — not its head, as with humans. The rover's brain comes in the form of computers, which are used to control all of the rover's motors, instruments, and communications between Earth and Mars. The neck and head of the rover are used to support two types of "eyes" or cameras: Pancams and Navcams.

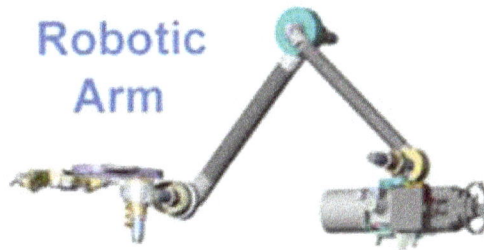

Robotic Arm

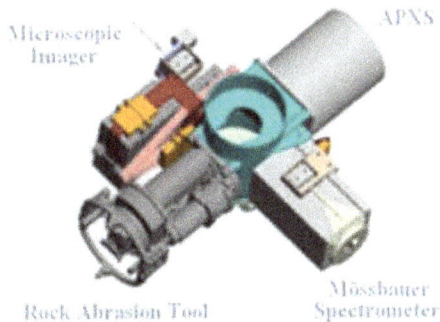

Specifics of a robotic arm.

Pancams are used to collect a panoramic view of the Martian terrain. Navcams take black and white pictures to capture 3-D imagery of the surface. This information is then sent to the brains of the rover to help navigate the rover and avoid obstacles. There are also Hazcams (Hazardous Avoidance Cameras) placed on the front and back of the rovers also to prevent the rovers from crashing into any unexpected obstacles. The rovers operate off of batteries which are recharged from solar panels on top of the rover.

Interestingly, the rovers only have one arm, but it has a wide range of motion and holds four different instruments to inspect and analyze the Martian terrain. Engineers designed the arm with five degrees of freedom — which gives it such wide range of motion. A degree of freedom can be illustrated with joints of your body. Your elbows and knees have one degree of freedom because you can only move them forward or backward. Your wrist has three degrees of freedom because you can move your wrist up and down, side to side, and rotate it clockwise and counterclockwise. Therefore, five degrees of freedom means that the rover's arm can move in nearly any direction/any degrees of range: up or down, front to back, side to side, side and up, front and down, etc.

Once the arm is in place, the four instruments can analyze the surface of Mars. The Microscopic Imager is a combination of a microscope and a camera that provides small-scale features of rocks and soil. Its importance is to help understand the properties of the rocks and soils and to help identify if water existed on Mars. Since most of the rocks and soil on Mars contain iron, the Mössbauer Spectrometer was designed to investigate iron-bearing minerals and analyze their composition as well as magnetic properties. The Alpha Particle X-Ray Spectrometer (APXS) is designed to study x-rays emitted by the rocks and soils to determine their elemental chemistry. Both the Mössbauer Spectrometer and APXS take about 10-12 hours to take measurements. The Rock Abrasion Tool (RAT) is a powerful grinder that can create holes 2-inches in diameter and .2-inches deep. Once a hole is created, the other instruments are used on the interior of the rock. This distinction is important because the rock's interior may be drastically different than its exterior, and the instruments reveal how the rock was formed and the environmental conditions in which it was altered.

Service Robot

Service robot is a robot that performs useful tasks for humans or equipment excluding industrial automation applications.

Robots require a degree of autonomy which is the ability to perform intended tasks based on current state and sensing, without human intervention. For service robots this ranges from partial autonomy - including human robot interaction - to full autonomy - without active human robot intervention. The IFR statistics for service robots therefore include systems based on some degree of human robot interaction or even full tele-operation as well as fully autonomous systems.

Professional Service Robots

Professional service robots are a type of robot typically considered for use outside of a manufacturing facility within a professional setting. While industrial robots automate manufacturing tasks, professional service robots, which vary greatly in form and function, automate menial, dangerous, time-consuming, or repetitive tasks, effectively freeing human workers to perform more cognitive functions.

Most professional service robots are semi-autonomous or fully autonomous robots with some form of mobility. There are service robots that are intended to interact with people, typically deployed in a retail, hospitality, healthcare, warehouse or fulfillment setting. Others are deployed in more rugged settings, such as in space and defense, agricultural applications, and demolition, to automate dangerous or laborious tasks.

Alongside the continued advances in vision systems and technologies, motion control and motors, artificial intelligence and machine learning, as well as cloud computing, the capabilities and applications of professional service robots are expanding at an exponentially faster rate than ever before.

Benefits of Professional Service Robots

Professional service robots automate many different tasks in a diverse set of applications, but the benefits they bring are typically the same. Regardless of industry or application, businesses choose to automate processes for the same reasons: safety, efficiency and productivity.

Safety

Safety is a primary consideration when businesses choose to deploy professional service robots. Rather than replacing human workers, robots can handle dangerous tasks while humans focus on more cognitive-oriented tasks that keep them away from dangerous situations. For example, professional service robots used in the defense sector aim to protect soldiers from harm during combat. Similarly, demolition robots are used to keep workers away from structural or nuclear dangers during the demolition process. Professional service robots can safely go where human workers cannot.

Efficiency

In addition to safety, improving efficiency is another primary consideration for businesses that use professional service robots. Efficiency is typically gained through impeccable levels of robot uptime and reduced operating costs. Inspection robots and industrial cleaning robots operate with very little downtime, allowing them to provide more comprehensive coverage of large infrastructures. Logistics robots can transport a high volume of products while lowering labor costs, too.

Productivity

Like all robotic automation, professional service robots often open the door to greater data collection and analysis for ongoing optimization of all operations. Agriculture robots can transmit important visual data to monitor the health of crops and livestock.

Customer service robots can even track customer behavior during face-to-face encounters to gain a deeper understanding of consumer motivations and desires. Essential operational data such as this is much more difficult or impossible to collect through manual processes.

In the end, professional service robots all contribute to productivity in one form or another. Some may excel at safety, while others may excel at data collection or efficiency, but the end result is almost always a major increase in productivity. The productivity gains of these robots help businesses of all kinds justify the investment in professional service robots.

Snakebot

Snakes are unique creatures in that their bodies allow them to get into the cracks and crevices of the world that most other creatures cannot. Lacking rigid skeletons and extremities, snakes can contort their bodies in order to get into tiny holes, wrap around tree branches and slither over otherwise unmanageable rocks. These serpentine qualities are the inspiration for a new type of robotic, interplanetary probe, called a snakebot, being developed by engineers at NASA's Ames Research Center.

Since 1964, NASA has sent 10 robotic explorers to fly by, orbit or rove around Mars, but snakebots will give scientists an unprecedented look at the Martian landscape. Snakebots, , will be able to dig into the loose soil of Mars and burrow down to depths that other robotic probes can't get to. They can slither into the cracks of the planet's surface. A snakebot could navigate over rough, steep terrain where a wheeled robotic rover would likely get stuck or topple.

Snakebots are expected to be more durable and cheaper than any probe that has ever been sent to investigate a planet.

Snakebots are unlike any robotic probe ever to be used for space missions. In order for a robot to mimic the movements of a biological snake, some special design features have to be used. NASA's snakebots are a model of the polybot. Polybots are robots that are able to change their shape in order to perform a variety of tasks. Snakebots will slither and dig underneath the soil for geological surveying, or coil up to carry tools for construction in space.

The main body of a snakebot consists of about 30, identical, hinge-like modules that are linked together in a chain. These modules are connected by a central spine and work together to perform various functions. The snakebot frame will be constructed out of a polycarbonate material and covered by an artificial skin to protect it from the Martian elements. Here's a closer look at a snakebot's architecture and individual modules:

- Electronics: Each snakebot will have a central computer, possibly located in the snakebot's head, that works in conjunction with smaller computers in each module. Wires will connect each module to its neighboring modules, creating a network of modules that work together as a unit. The wiring will also carry communications and power to and from the computer brain.

- Microcontrollers: These tiny computers will interpret signals from the main computer to control movement. In later models, they may be connected to a set of sensors to provide reflexes.

- Sensors: In later models, strain sensors may be added to the robot's metal-rib frame. These sensors will indicate if the snake is in contact with anything, where it's touching it and how hard the contact is.

- Motors: Two servomotors, which are like off-the-shelf hobby motors, will be used to move the various parts in each module. Each motor will be activated by a signal from the main processor.

- Wheels: Each module will be equipped with one wheel. The wheel won't be wholly responsible for transporting the snakebot -- it will only used to ease movement.

- Gears: Working in conjunction with the electronics, the gears will allow for movement of the hinges. This will give the snake the ability to coil, side-wind and inch-worm its way across the ground or wrap around objects.

- Camera: Small cameras attached to the snakebots will give NASA a never-be-fore-seen view of the red planet.

- Connecting Rods: When one section begins to move, these ball-jointed connecting rods will pull and activate the section next to it.

An up-close look at the snakebot modules

Snakebots will be able to limit the weight of the spacecraft ferrying them to space. The snake-like design allows them to perform many tasks without a lot of extra equipment. "One of the many advantages of the snake-based design is that the robot is field-repairable," NASA engineer Gary Haith says. "We can include a bunch of identical spare modules with the snake on a space mission, and then we can fix the snakebot much easier than a regular robot that needs specific parts".

Unlike past robotic probes, snakebot will be very cheap. In contrast to the $135-million Mars Odyssey that was launched on April 7, 2001, snakebots will probably cost only a few hundred dollars each. In fact, the cost of the snakebot is so low that one researcher says there is a possibility of developing a toy version.

Surgical Robot

A surgical robot is a self-powered, computer-controlled device that can be programmed to aid in the positioning and manipulation of surgical instruments, enabling the surgeon to carry out more complex tasks. Systems currently in use are not intended to act independently from human surgeons or to replace them. Instead, these machines act as remote extensions completely governed by the surgeon and thus are best described as master-slave manipulators. Two master-slave systems have received approval by the US Food and Drug Administration (FDA) and are in use– the da Vinci Surgical System and the ZEUS. Each system has 2 basic components linked together through data cables and a computer .

- The surgeon's master console is the robot's user interface that provides the master surgeon with the following functions:

  - A 3-dimensional view of the surgical field relayed from an endoscopic camera inside the patients body in control of the robot that creates a sense of being "immersed" into the surgical field.

  - Master manipulators, which are handles or joysticks that the surgeon uses to make surgical movements that are then translated into real-time movements of the slave manipulators docked on the patient. Motion scaling (conversion of large natural movements to ultraprecise micro movements) and tremor filtering increase accuracy and precision of the surgeon's movements.

  - A control panel to adjust other functions, such as focusing of the camera, motion scaling, and accessory units.

- Patient-side slave robotic manipulators are robotic arms that manipulate the surgical instruments and the camera through laparoscopic ports connected to the patient's body. The da Vinci system handles surgical instruments with micro articulations near the tip (EndoWrist) that can duplicate motions of the human wrist, including rotation (7 degrees of freedom, ie, the greatest possible motion around a joint).

Clinical Applications of Robotic Surgery

Robotic surgery has successfully addressed the limitations of traditional laparoscopic and thoracoscopic surgery, thus allowing completion of complex and advanced surgical procedures with increased precision in a minimally invasive approach. In contrast to the awkward positions that are required for laparoscopic surgery, the surgeon is seated comfortably on the robotic control consol, an arrangement that reduces the surgeon's physical burden. Instead of the flat, 2-dimensional image that is obtained through the regular laparoscopic camera, the surgeon receives a 3-dimensional view that enhances depth perception; camera motion is steady and conveniently controlled by the operating surgeon via voice-activated or manual master controls. Also, manipulation of robotic arm instruments improves range of motion compared with traditional laparoscopic instruments, thus allowing the surgeon to perform more complex surgical movements.

Table: Laparoscopic Limitations/Robotic Solutions

Laparoscopic Problems/Limitations	Robotic Surgery Solutions/Potential
Two-dimensional vision of surgical field displayed on the monitor impairs depth perception	Binocular systems and polarizing filters create 3-dimensional view of the field
Movements are counterintuitive (ie, moving the instrument to the right appears to the left on the screen due to mirror-image effect)	Movements are intuitive (ie, moving the control to the right produces a movement to the right on the viewer)
Unstable camera held by an assistant	Surgeon controls camera held in position by robotic arm, allowing solo surgery
Diminished degrees of freedom of straight laparoscopic instruments	Microwrists near the tip that mimic the motion of the human wrist
Surgeon forced to adopt uncomfortable postures during operation	Superior operative ergonomics: surgeon comfortably seated on the control console
Steep learning curve	Shorter learning curve

In a relatively short time, robotic procedures spanning the whole spectrum of surgery have been successfully executed. Initial results show that mortality, morbidity, and hospital stay compare favorably to conventional laparoscopic operations. However, only a limited number of randomized, prospective studies that compare outcomes of robotic techniques with conventional methods exist. More procedure-specific, randomized trials need to be performed before robotic surgery can find its way into everyday surgical practice.

Table: Clinical Applications of Robotic Surgery

Field	Operations Performed via Robotic Surgery
Robotic gastrointestinal surgery	1997: Himpens– first robotic cholecystectomy
Robotic urologic surgery	Antireflux operations, Heller's myotomy, gastric bypass, gastrojejunostomy, esophojectomy, gastric banding colectomy, splenectomy, adrenalectomy, and pancreatic resection reported to date
Robotic gynecologic surger	Radical robotic prostatectomy is the most common operation performed robotically and is gaining widespread recognition in the United States and EuropeNephrectomy and pelvic lymph node dissection also reported
Robotic cardiothoracic surgery	Robotic hysterectomy, salpingo-oophorectomy, and microsurgical fallopian tube reanastomosis
	Surgical robots allow cardiothoracic surgeons to perform complex cardiothoracic procedures while avoiding the significant morbidity of sternotomy and thoracotomyHundreds of robotic coronary bypasses have been performed to dateMitral valve repairs, atrial spetal defect repair, pericardiectomy, lobectomies, and tumor enucleations
Robotic oncologic surgery	Esophageal tumors, gastric cancer, colon cancer, thymoma, and retromediastinal tumors
Robotic pediatric surgery	Pyeloplasty for ureteropelvic junction obstruction, antireflux procedures for gastroesophageal reflux disease, and pediatric congenital heart diseases, such as ligation of patent ductus arteriosus

Limitations of Robotic Surgery

Although rapidly developing, robotic surgical technology has not achieved its full potential owing to a few limitations. Cost-effectiveness is a major issue; 2 recent studies comparing robotic procedures with conventional operations showed that although the absolute cost for robotic operations was higher, the major part of the increased cost was attributed to the initial cost of purchasing the robot (estimated at $1,200,000) and yearly maintenance ($100,000). Both factors are expected to decrease as robotic systems gain more widespread acceptance. However, it is conceivable that further technical advances may at first drive prices even higher. Decreasing operative time and hospital stay will also contribute to the cost-effectiveness of robotic surgery.

Other drawbacks to robotic surgery include the bulkiness of the robotic equipment currently in use. Lack of tactile and force feedback to the surgeon is another major problem,for which haptics (ie, systems that recreate the "feel" of tissues through force feedback) offers a promising, although as yet unrealized, solution.

Robotic Surgery Procedures

Robotic Colorectal Surgery

During a robotic colectomy, surgeons remove cancerous portions of the colon and rectum, as well as benign tumors and polyps. A robot-assisted approach provides surgeons with the tools to more easily connect the two ends of the colon after the cancer has been removed. The procedure can be completed with a few tiny incisions, rather than the one long incision used in traditional open colon surgery.

Robotic surgery allows surgeons to perform complex rectal cancer surgery, which had been extremely challenging, in a minimally invasive manner. The robot provides improved visualization of the surgical site through 3D magnification, enhanced dexterity for manipulation and dissection of tissue, and greater precision.

The robotic procedure allows surgeons to finely dissect cancers of the rectum while possibly reducing nerve injury. A recent study has shown that surgeries using the robot are less likely to require conversion to an open procedure than colorectal procedures performed laparoscopically.

Single-site Robotic Gallbladder Surgery

Treatment for gallbladder disease generally includes lifestyle and dietary changes and medication. However, in some cases, doctors recommend cholecystectomy, which is surgery to remove the gallbladder.

Although minimally invasive cholecystectomy has been used for several years, a new surgical refinement, in which only one incision, in the navel (belly button), is made to perform the surgery. Doctors at Summerl in Hospital use the da Vinci Xi Surgical System, a refinement to the standard da Vinci system. Because there is only a single, small incision, patients often experience less trauma to the body and less blood loss during the procedure surgery. This means they may recover more quickly and return to work and daily activities faster than they would with conventional surgery.

References

- What-are-autonomous-robots: robots.com, Retrieved 13 July 2018
- Different-types-of-autonomous-robots-and-real-time-applications: elprocus.com, Retrieved 10 May 2018
- Android-humanoid-robot: searchenterpriseai.techtarget.com, Retrieved 14 June 2018
- The-most-life-life-robots-ever-created: futurism.com, Retrieved 30 March 2018
- What-are-the-main-parts-of-an-industrial-robot: robots.com, Retrieved 09 July 2018
- Mobile-robotics-33065: techopedia.com, Retrieved 10 April 2018
- Nanobots-kill-drill-cancer-cells-60-seconds: weforum.org, Retrieved 19 June 2018

3

Components of a Robot

Some of the significant components of a robot are the power source, actuator, electric motor, sensors, etc. This chapter explores the diverse components integral to the design of robots that are important for the effective function of a robot. It also discusses in detail about robotic non-destructive functioning, forward chaining, servomechanism, etc

Actuator

Actuators have become a crucial part of the automated system, as they help with controlling equipment using hydraulic, pneumatic or in some of the cases electronic signals. Linear actuators are quite common as they form the bulk of applications used in the control of equipment moving along a lateral axis. However it is also important to note that there are other types of actuators that are found in robotic arms. One of the actuators types is referred to as a synchronous actuator and relates to the fact that it has a motor which rotates synchronously with the oscillating current. The brushless DC servo is also a common one and like the name suggests it uses a DC current and does not have brushes.

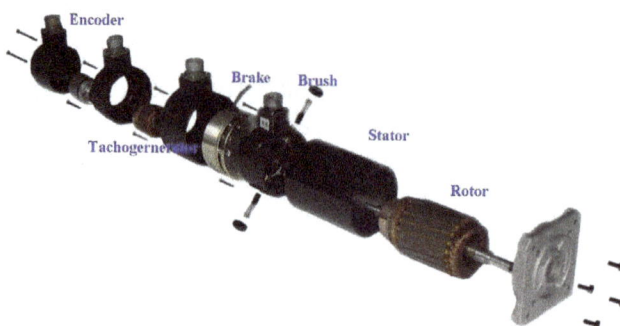

Instead it has permanent magnet poles on the rotor which are attracted to the magnetism of the rotating poles. They tend to last longer and give off no noise along with reduced interference from electromagnetism.

The stepper is also brushless and moves in small but discrete steps. It can be used with digital and analog signals. The brushed DC servo on other hand has brushes/commutators which are found on a permanent magnet or a wound stator. The asynchronous actuator on the other hand is designed to slip for it to generate torque. Traction motor is an electric motor that is common cars for driving the wheels. AC servo motors are two phased just like the driving current and are crucial where fast and accurate responses are required.

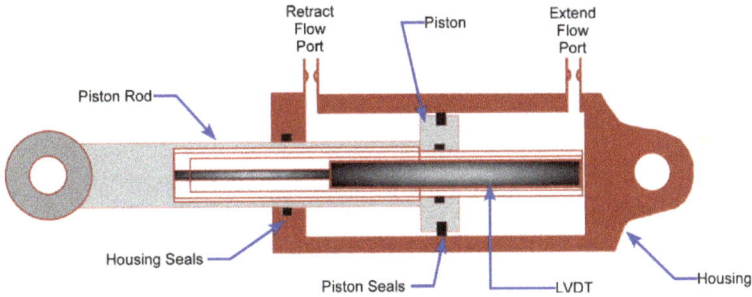

Types of Robot Actuators

Pneumatic Actuator

These actuators use the principle of Pneumatics (where pressurized air/gas) to create motion (or perform work). Pneumatic actuators can be used to produce both rotational motion and linear motion. Pneumatic Actuator The basic design includes a Cylinder, a Piston and few supporting components like valves, springs, stoppers etc., Compressed Air or pressurized gas is filled into the cylinder and the compressed air tries to expand to reach atmospheric pressure. This expansion is forced towards a piston or any other mechanical device which makes an attached object move. Depending on the design, it can either create rotational motion or linear motion. Most air compressors and air pumps use this principle due to its simplistic design. Image shows single acting cylinder where a spring pushes the piston back to its base position. In a double acting cylinder, another valve on other end of the cylinder pushes the piston back to base position.

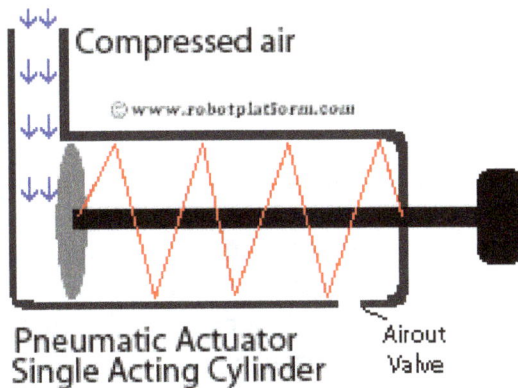

Pneumatic actuators are mainly used for systems which require quick and accurate response. These actuators are clean, make less noise and relatively compact in their design.

Air Muscles: To replicate muscles in robots, Air muscles (also known as pneumatic air muscles - PAM) are used. PAM's generally consist of a rubber bladder covered by a braided fiber mesh. When pressurized gas/air is inflated, it expands radically and when deflated, it contracts. Air muscles are inexpensive, light weight, exhibit a phenomenal strength to weight ratio, easier to build, flexible compared to other electric and hydraulic actuators, durable, safe, and also easy to use under water.

Hydraulic Actuator

According to Blaise Pascal, when there is an increase in pressure at any point in a confined incompressible fluid, then there is an equal increase at every point in the container. Hydraulic actuators are designed based on this principle (Pascal's law).

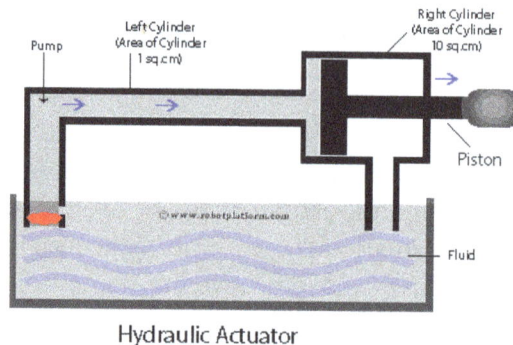

Hydraulic Actuator

To understand how hydraulic actuators work, let us take an example of two cylinders, connected together, as shown in the image. Suppose one cylinder has a cross-section area of 1 square centimeter and the second one has a cross-section area of 10 square centimeters. If the cylinders are filled with incompressible fluid and 1 unit of pressure is applied to the left cylinder pushing the pump (actually liquid) by 10 centimeter, then the resulting force acts on the right cylinder pushing the piston by 1 centimeter, but with a force of 10 units. This means applying 1 unit of force produces 10 units of force on the other side.

Hydraulic actuators are majorly used for systems which require very large force, but not very restrictive on positioning and accuracy.

Hydraulic Cylinder: Generally referred as linear hydraulic motor is used when a robot requires linear force. These actuators are powered by hydraulic fluids. When hydraulic pressure acts on it, a piston connected to a piston rod within the cylinder moves back and forth creating linear motion. Hydraulic cylinders can also convert hydraulic pressure into rotation creating a rotary actuator, based on the mechanical design.

Piezoelectric Actuators

These actuators use piezoelectric effect to create motion. For those who are not aware of piezoelectric (piezo means press, or apply pressure) effect, it is a charge that accumulates in certain ceramic materials and crystals (like quartz, tourmaline etc.) when a mechanical stress is applied. This is known as piezoelectric effect where deformation of the material creates a voltage difference.

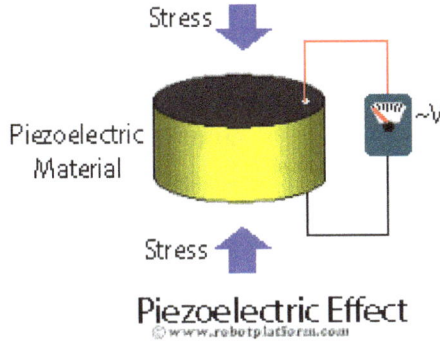

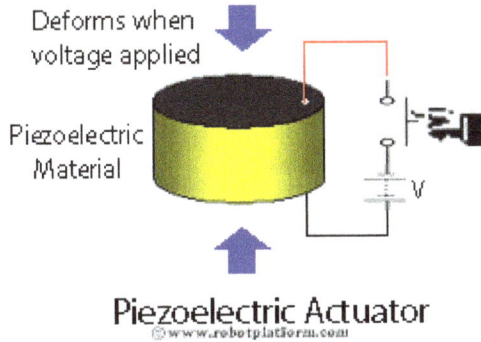

Piezoelectric Effect Piezoelectric Actuator

When an electricity flows through a piezoelectric material, it creates a physical deformation which is proportional to the applied electric field, known as indirect piezoelectric effect. This precise deformation can be used to position objects with extreme accuracy, almost at μm accuracy.

Piezoelectric actuators are used in Loudspeakers, Piezoelectric Motors, acceleration sensors, vibration sensors etc., and can be used to create either rotational or linear motion. The strokes of these actuators can also be amplified if required, because direct strokes from these peizoelectric actuators are generally less than 100 μm.

Ultrasonic Piezoelectric Actuators

Similar to Piezoelectric actuators, Ultrasonic Piezoelectric actuators work on the principle of piezoelectric effect. A travelling wave excites a stator surface (material exhibiting piezoelectric effect) which behaves like an elastic ring and produces elliptical motion at the interface of the rotor, which propels the rotor and the drive shaft connected to it.

Ultrasonic Piezoelectric actuator can induce either rotational motion or linear motion.

These actuators provide extremely precise movement with good torque to size ratio, and used in most camera autofocus lenses and watch motors.

Delta Robot

Delta robots are a type of parallel robot, meaning that their motors and legs are connected mechanically in parallel, with each leg supporting only itself. In contrast, SCARA, 6-axis, and Cartesian styles are serial robots, meaning that each arm has to carry the next arm beyond it. For example, in an X-Y-Z Cartesian robot, the Y axis must support the Z axis, and the X axis must support both the Y and Z axes.

Delta robots operate on the basis of three parallelograms connected to a stationary base above and a moving platform below. The base is essentially an equilateral triangle, with three legs connected to it—one on each side of the triangle—each moved by a small motor. These legs are jointed (think of knees), with the lower sections consisting of two bars in parallel, and the lower end of each bar connected to the moving platform, which is also an equilateral triangle. These lower sections of the legs make up the three parallelograms that control the platform, constraining it to X, Y, and Z motion, parallel to the base.

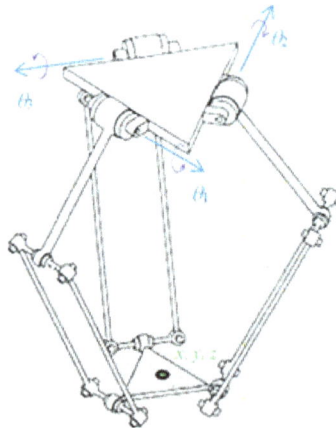

From a mathematics standpoint, the kinematics of a delta robot are relatively straightforward. Three motors manipulate three parallelograms to move a platform in the X, Y, and Z directions.

A fourth axis can be included for rotation, either by attaching a rotary actuator directly to the platform, or by adding a leg that extends from the center of the base to the center of the platform, with a rotating joint on the end. (Note that this leg must be telescoping, as its length will change as the three primary legs change position).

Design Advantages of Delta Robots

Delta robots dominate the field when it comes to speed and acceleration, because only their legs and platform move (along with the payload, of course). This means that they have much less moving mass and inertia than SCARA, 6-axis, or even Cartesian robots, and can achieve accelerations in the range of 150 m/s2 and maximum speeds up to 10 m/s. To further capitalize on this design benefit, the legs and platform can be made from lightweight materials, such as aluminum or carbon fiber.

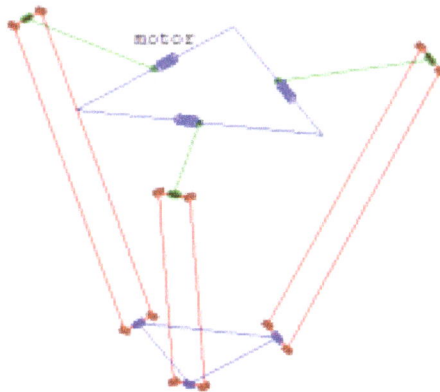

Three parallelograms (in red), are connected to the stationary base and moving platform (in blue).

Unlike serial robots, in which the errors in each arm accumulate, in the parallel robot design, the errors in each leg are averaged. This gives delta robots the ability to achieve sub-millimeter repeatability. The parallel design also provides high stiffness, because the legs work together against external forces and torques. (Note that the stiffness of the legs, which is influenced by their material and size, also has an effect on the robot's stiffness).

Being mounted above the workspace, delta robots have no interferences in their work area, meaning there are no "dead zones," or zones that the end effector cannot reach within the defined work envelope. Overhead mounting also makes delta robots easy to adapt for harsh or challenging environments. With the major electromechanical parts mounted above the work area, they can be shielded from harsh chemicals or debris, or enclosed to make them clean-room or wash-down compatible. And the legs can be made of virtually any material that is suitable for the environment.

Applications then and Now

Delta robots were introduced for industrial applications in the early 1990's, with the first robots being employed in the packaging industry. As vision systems and force sensors were integrated, delta robots began to move beyond simple pick-and-place operations, and by the late 1990's, their applications had expanded into the electronics and assembly markets.

More recently, delta robots are being deployed in high-speed assembly and light machining applications. And new industries, such as 3D printing, are taking advantage of their relative mechanical simplicity and high-speed capabilities. Now, delta robots can be found in virtually any application that requires the ability to achieve extremely fast speeds and accelerations with light payloads.

End-effector

An End Effector is considered as the "hand of a robot." It is one of the important devices in a robot. It integrates an arm and a wrist, which helps it to perform several functions like material handling, pick and place, machine loading and unloading, etc. It is also known as EOA (End of Arm Tooling), Robotic Peripherals, and more.

Based on the performance of work tasks, the robot end effector can be described into two important categories such as:

- Grippers, and
- Tools.

Grippers

The grippers are mainly used in a robot for grasping / holding an object. The grasped objects will be moved to the preferred place with the help of a robot. The grippers are capable of carrying work parts, bottles, tools, and so on. It can work by means of magnetic, mechanical, vacuum cups, etc.

The grippers can also be sorted as:

- Single gripper,
- Double grippers, and
- Multiple grippers.

If a robot wrist contains one grasping device, then it is described as Single Gripper. Similarly, a robot wrist with two grasping devices will be termed as Double Grippers. It will be very much suitable for machine loading and unloading operations. Moreover, it is better than single gripper, because it finishes plenty of works in a quick time. The Multiple Grippers incorporates more than two grasping devices in a robot wrist. The double grippers seem to be the best one than the multiple grippers. The reason is that the multiple grasping mechanisms are pricing very high. The grippers can be further classified into internal gripper and external gripper.

Other types of grippers are:

- Magnetic Grippers,

- Vacuum Cups, and

- Adhesive Grippers.

A tool is equipped in the robot for carrying out several operations on the work parts instead of grasping it. A tool acts as an end effector when it is attached directly to the robot's wrist. In some applications, there will be a need for multi-tool task, and changing the tool all the time from the robot wrist will be highly difficult. As a result, a gripper is used in this process to grasp and manipulate the tool. It certainly helps the robot to handle several tools in an operation, and thus makes the multi-tooling function possible. Moreover, the time taken to change the tool is very low.

For example: In a deburring operation, various sizes of deburring tool will be required to hold for reaching every surface of the work part. Here, the tool is equipped in the gripper for quick exchange from one tool to another.

Examples for Tools used as end Effectors

In robot applications, the most commonly used three tools as end effectors are listed below:

- Spot welding tools

- Spray painting nozzle

- Arc welding torch

The rotating spindle for routing, drilling, grinding, and wire brushing operations also comes under this category. Some other examples of tools used as end effectors are liquid cement applicators, water jet cutting tool, and heating torches.

In all the above examples, the actuation of a tool must be coordinated by an industrial robot. An example of this case includes the spot welding operation in which the robot must control the actuation as a part of its process. The control of this process is very similar to the opening and closing of a mechanical gripper.

Forward Chaining

Often rule-based systems work from just a few facts but are capable of reaching many possible conclusions. Examples are "sensory" and "diagnosis" expert systems, like those that identify an object from a description of what you see at a distance, or those that tell you what to do when your car breaks down from a description of what isn't working. For these, it often makes more sense to start with the facts and reason to goals, what is known as forward chaining or data-directed computation or modus ponens reasoning.

As an example, take the rule

   a :- b, c.

and suppose b and c are facts. Then we can conclude a is a fact, and add it to the facts we have. This is called modus ponens inference in logic. There's no query, no goals; we just use the facts we know to infer some new fact. (The built-in predicate asserta can help implement this too).

To use modus ponens as the basis of a control structure, take the facts in order. For each fact, find all rules whose right sides contain a predicate expression that can be matched to it. (We can index predicate names mentioned on right sides to speed this up.) Now "cross out" the predicate expressions matched in the rules; that is, create new rules like the old rules except without these expressions. But wherever a fact matched the last expression on the right side of some rule, the left side of the rule has been proved a new fact, after substituting in any bindings made on the right side; so cache that new fact in the database. The last paragraph gave an example. For another, consider:

   a(X,3) :- b(X).

Then if b(20) is a fact, modus ponens would conclude a(20,3) is a fact.

It matters where we put a new fact among the others, since the facts are pursued in order. Usually it's good to put the new fact in front of those not yet considered, a most-recent-fact or focus-of-attention idea. So the fact just proved will be the next fact examined; the fact list then is much like a stack (last thing "in" is the first thing "out"). This might be good if we want to reach some particular conclusion as fast as possible. But, on the other hand, if we want to systematically find every conclusion possible from the facts, we should put new conclusions at the end of the set of facts; the fact list is then like a queue (last thing "in" is the last thing "out").

Any nots in rules require special handling. Since we want to follow the closed-world assumption for forward chaining too (it's simplest), we want not to mean "it can't be proved". So forward chaining must first assume all nots to be false, prove all possible facts, and only then consider as true those nots whose arguments are not now facts. Those nots may then prove new facts with new consequences. (To avoid such awkwardness, some artificial-intelligence systems let you state certain facts to be false, but this creates its own complications).

Here's the formal algorithm for (pure) forward chaining:

1. Mark all facts as unused.

2. Until no more unused facts remain, pick the first-listed one; call it F. "Pursue" it:

   (a) For each rule R that can match F with a predicate expression on its right side, ignoring nots, and for each such match in the rule (there can be multiple match locations when variables are involved):

(i) Create a new rule just like R except with the expression matching F removed. If variables had to be bound to make the match, substitute these bindings for the bound variables in the rule.

(ii) If you've now removed the entire right side of rule R, you've proved a fact: the current left side. Enter that left side into the list of facts, and mark it "unused". (The focus-of-attention approach here puts the new fact in front of other unused facts.) Eliminate from further consideration all rules whose left sides are equivalent to (not just match able to) the fact proved.

(iii) Otherwise, if there's still some right side remaining, put the new simplified rule in front of the old rule. Cross out the old rule if it is now redundant. It is redundant if the old rule always succeeds whenever the new rule succeeds, which is true when no variables were bound to make the match.

(b) Mark F as "used".

3. Create a new fact list consisting of every not expression mentioned in rules whose argument does not match any used fact. Mark all these as "unused", and redo step 2.

A Forward Chaining Example

Let's take an example of forward chaining with focus-of-attention handling of new facts.

/* R1 */ goal1 :- fact1.

/* R2 */ goal1 :- a, b.

/* R3 */ goal2(X) :- c(X).

/* R4 */ a :- not(d).

/* R5 */ b :- d.

/* R6 */ b :- e.

/* R7 */ c(2) :- not(e).

/* R8 */ d :- fact2, fact3.

/* R9 */ e :- fact2, fact4.

Suppose the rules are taken in the given order; and that as before, only fact2 and fact3 are true, in that order.

1. We start with fact2, and find the matching predicate expressions R8 and R9. This gives the new rules

> /* R10 */ d :- fact3.

> /* R11 */ e :- fact4.

Rules R8 and R9 are now redundant since no variables were bound, and R8 and R9 can be eliminated.

2. No new facts were discovered, so we pursue next fact3. This matches an expression in R10. So now R10 succeeds, and the new fact d is put in front of any remaining unused facts (though there aren't any now). R10 can be eliminated.

3. We pursue fact d, and find that rule R5 mentions it in its right side. (R4 mentions it too, but as not(d), and we're saving nots for last.) Matching R5 gives the new fact b. Rules R5 and R6 can now be eliminated.

4. Fact b matches in R2, giving

> /* R12 */ goal1 :- a.

and rule R2 can be eliminated. The current set of rules is:

> /* R1 */ goal1 :- fact1.

> /* R3 */ goal2(X) :- c(X).

> /* R4 */ a :- not(d).

> /* R7 */ c(2) :- not(e).

> /* R11 */ e :- fact4.

> /* R12 */ goal1 :- a.

5. We have no more facts to pursue. But we're not done yet, since R4 and R7 have nots.

6. Fact d is true, so rule R4 can't ever succeed. But fact e has not been proved. Hence add not(e) to the list of facts.

7. This matches the right side of R7. Hence c(2) is a fact too. Eliminate R7.

8. This matches the only expression on the right side of R3, when X=2, and hence goal2(2) is a fact. We can't eliminate R3 now because we had to bind a variable to make the match, and we can still use R3 for other values of X.

9. That's everything we can conclude.

Manipulator

Industry-specific robots perform several tasks such as picking and placing objects, movement adapted from observing how similar manual tasks are handled by a fully-functioning human arm. Such robotic arms are also known as robotic manipulators. These manipulators were originally used for applications with respect to bio-hazardous or radioactive materials or for use in inaccessible places.

A series of sliding or jointed segments are put together to form an arm-like manipulator that is capable of automatically moving objects within a given number of degrees of freedom. Every commercial robot manipulator includes a controller and a manipulator arm. The performance of the manipulator depends on its speed, payload weight and precision. However, the reach of its end-effectors, the overall working space and the orientation of the work is determined by the structure of the manipulator.

Kinematics of a Robotic Manipulator

A robot manipulator is constructed using rigid links connected by joints with one fixed end and one free end to perform a given task (e.g., to move a box from one location to the next). The joints to this robotic manipulator are the movable components, which enables relative motion between the adjoining links. There are also two linear joints to this robotic manipulator that ensure non-rotational motion between the links, and three rotary type joints that ensure relative rotational motion between the adjacent links.

The manipulator can be divided into two parts, each having different functions:

Arm and Body – The arm and body of the robot consists of three joints connected together by large links. They can be used to move and place objects or tools within the work space.

Wrist – The function of the wrist is to arrange the objects or tools at the work space. The structural characteristics of the robotic wrist includes two or three compact joints.

Robotic Manipulator Arm Configuration

- Cartesian geometry arm – This arm employs prismatic joints to reach any position within its rectangular workspace by using Cartesian motions of the links.

- Cylindrical geometry arm – This arm is formed by the replacement of the waist joint of the Cartesian arm with a revolute joint. It can be extended to any point within its cylindrical workspace by using a combination of translation and rotation.

- Polar/spherical geometry arm – When a shoulder joint of the Cartesian arm is replaced by a revolute joint, a polar geometry arm is formed. The positions of end-effectors of this arm are described using polar coordinates.

- Articulated/revolute geometry arm - Replacing the elbow joint of the Cartesian arm with the revolute joint forms an articulated arm that works in a complex thick-walled spherical shell.

- Selective compliance automatic robot arm (SCARA) – This arm has two revolute joints in a horizontal plane, which allow the arm to extend within a horizontal planar workspace. The TH650A SCARA Robot by TM Robotics is a great example to demonstrate pick and place functionality of robotic manipulators.

Wrist Configuration

The two main types of wrist design include:

- Roll-pitch-roll or spherical wrist

- Pitch-yaw-roll.

The spherical wrist is more common because of its mechanically simpler design. It has 6 degrees of freedom, and consists of a Hooke shoulder joint followed by a rotary elbow joint.

Applications

Some of the major applications of robotic manipulator are discussed below:

- Motion planning,

- Remote handling

- Micro-robots

- Humanoid robots

- Machine tools.

Parallel Manipulator

Parallel manipulators are widely popular recently even though conventional serial manipulators possess large workspace and dexterous maneuverability. The basic problems with serial one are their cantilever structure makes them susceptible to bending at high load and vibration at high speed leading to lack of precision and many other problems. The serial one are compared. Hence, in applications demanding high load carrying capacity and precise positioning, the parallel manipulators are the better alternatives and the last two decades points to the potential embedded in this structure that has not yet been fully exploited. Willard L. V. Polard designed and patented the first industrial parallel robot as shown in. The development of parallel manipulators can be dated back to the early 1960s, when Gough and Whitehall first devised a six-linear jack system for use as a universal tire testing machine. Later, Stewart developed a platform manipulator for use as a flight simulator. Since 1980, there has been an increasing interest in the development of parallel manipulators. This paper highlights the potential applications of parallel manipulators include mining machines, walking machines, both terrestrial and space applications including areas such as high speed manipulation, material handling, motion platforms, machine tools, medical fields, planetary exploration, satellite antennas, haptic devices, vehicle suspensions, variable-geometry trusses, cable-actuated cameras, and telescope positioning systems and pointing devices. More recently, they have been used in the development of high precision machine tools by many companies such as Giddings & Lewis, Ingersoll, Hexel, Geodetic and Toyoda, and others. The Hexapod machine tool [5,6] is one of the widely used parallel manipulators for various industries.

Classification of Parallel Manipulators

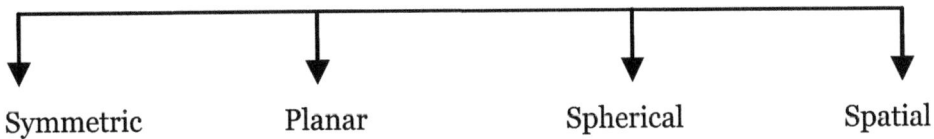

| Symmetric | Planar | Spherical | Spatial |

Symmetrical manipulators has number of limbs equals to number of degree of freedom, which is also equals to total numbers of loops. A planar parallel manipulator is formed when two or more planar kinematic chains act together on a common rigid platform. Now days, each leg of a planar parallel manipulator is replaced by a single wire, the manipulator is referred to as a planar wire-actuated (or wire-suspended) parallel manipulator. Spherical manipulators are just able to make the end effectors movement according to controlled spherical motions. Especially, spatial parallel manipulators with fewer degrees of motion than six, but more than three, attracted the attention of both, the researchers and users.

In the most of the research work, kinematic analysis is carried by either kinematic constraint equations, screw theory method, DH Parameters or using the concept of dual unit quaternion which is really a most efficient for representing screw displacements of lines for spatial manipulators. Using the inverse dynamics, the forces and torques of the actuated joints can be computed for closed loops configurations of parallel architecture. Newton Euler approach is relatively more economical compared to Euler Lagrange formulation for parallel as well as hybrid manipulators.

Industrial Applications

In 1942, a patent was issued to Willard L. V. pollard for his novel design of automatic spray painting. In fact, it was never built. It has five degree-of-freedom motion—three for the position of the tool head, and the other two for orientation. In 1954, Dr. Eric Gough, an employee of Donlop Rubber Co., England had developed a six degree of freedom—the first octahedral hexapod for universal Tyre Testing Machine. In 1965, Stewart published a paper in which he proposed a six degree of freedom parallel platform as a flight simulator. The Stewart platform comprises six pods, six spherical joints and six universal joints with 6 degree of freedom as shown in. Stewart platform has also been used for the Agile Eye (a spherical Parallel mechanism) —pointing devices developed by Gosselin and Hamel at robotics laboratory at Laval University, Canada, has low inertia and inherent stiffness, the mechanism can achieve angular velocities superior to 1000 deg/sec and angular accelerations greater than 20,000 deg/sec2, large- ly outperforming the human eye. An American engineer, Klaus Cappel is considered as a third pioneer in field of parallel robotics. He has developed an octahedral Hexa- pod Manipulator as motion simulator & was patented in 1967 as shown in F. Stewart Platform is also used for underground excavation device by Arai in 1991 and another application such as milling machines by Aronson in 1996.

3-DOF spherical parallel mechanism as shown in has interesting characteristics for practical ap- plications like worktable, a manipulator, a camera ori- enting device, a wrist or a motion simulator. Parallel Robots can offer many advantages for high-speed laser operations as shown due to their structural stiffness and limited moving masses with less power consumption.

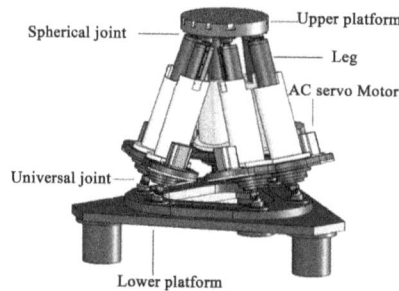

Typical Gough Stewart platform

First design of an octahedral hexapod for flight simulator (courtesy of Klaus Cappel)

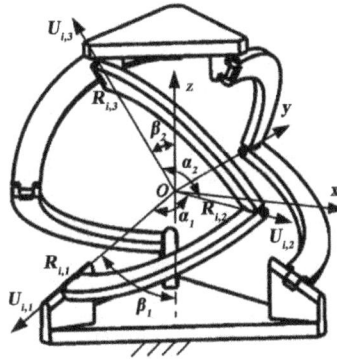

Three DOF spherical parallel manipulators

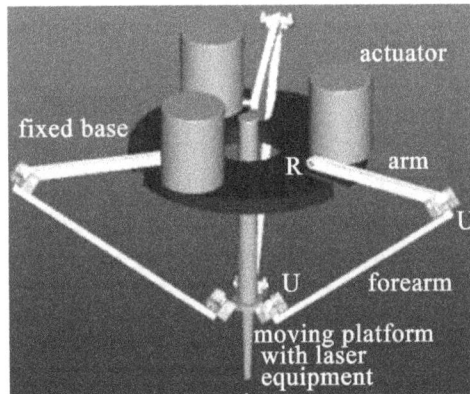

Three RUU kinematic chains usage in delta robot for laser cutting

depicts the parallel cube-manipulators have characteristics of no singularities in the workspace, simple form of forward kinematics and existence of a compliance center x = y = z = 0 with a added advantages of high stiffness and compactness used in the fields of micro-motion manipulators, remote center compliance (RCC) devices, assembly, planar kinematics machines and so on.

Remote Manipulator

Several stations have built in features that allow equipment, such as TRIUMPH cameras, to be deployed to specific locations in the reactor. These machines used to carry out this deployment are called manipulators and come in many shapes and sizes.

The size and shape of the manipulators is generally determine by the need to ensure that reactor gases do not escape whilst carrying out ISI – this is known as keeping containment. In order to achieve this most manipulators have a containment box or a sealing system and when in use are sealed onto the reactor penetration.

Manipulators are generally controlled from a dedicated console which contains joysticks, drivers, power supplies, camera monitors, computers and indicators. The power and signals to and from the manipulator are carried via umbilical which can be heavy and difficult to handle.

Peripheral Manipulator   Lower Boiler   Hoist Unit   Interstitial Manipulator
Annulus Manipulator

Example of Remote Manipulator

1. Hoist Unit

This is a very simple form of manipulator and basically consists of a winch housed inside a containment box. In order to keep containment, operations are carried out within the containment box which have built in gloves. These holes are known as gloveports. They can also be sealed off more permanently by use of gloveport valves or seals.

Hoist units are generally used to lower and raise equipment vertically in and out of the reactor core. These are generally used to inspect fuel and control rod channels as they can be easily positioned vertically above the channel.

To inspect components which are not vertically below the manipulator, a set of arms, similar to a crane davit, can be fitted inside the containment and used to lift the equipment sideways. From this new position the equipment can be lowered and raised. These arms are known as Elevating Guide Tubes (EGT) and can typically reach up to 2m.

Gloveport

2. Links Manipulators

In the majority of cases EGTs have insufficient reach to position the camera in the desired location. In these situations the cameras are deployed by links manipulators which have variable extensions up to 7m. The rigid extending arm is formed by a series of links, typically made from titanium. These are hinged at the top and can only bend in one direction. When deployed horizontally the hinges bend until the faces of adjoining links butt up against each other forming the arm. As the penetrations into the reactor are generally vertical, it is necessary to insert the links vertical then bend them around to the horizontal.

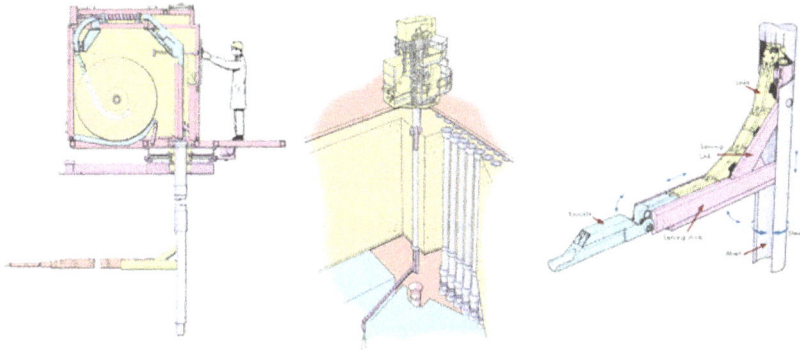

Examples of how a link manipulator functions

3. Extending Arm Manipulator

Another type of machine currently in use is the extending arm manipulator. In these devices the rigid arm is formed by extending telescopic sections. These manipulators are particularly useful where access can be gained horizontally, for example via the gas circulator penetrations.

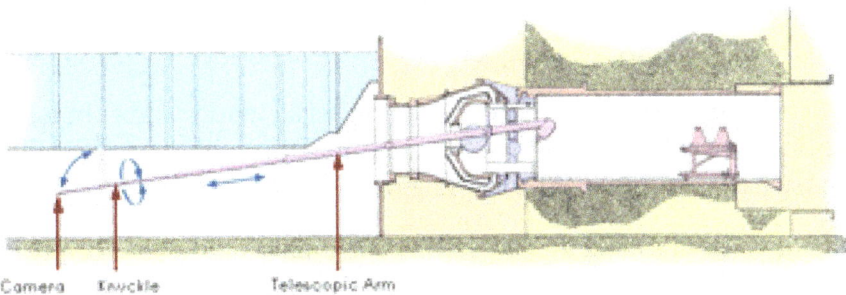

Camera   Knuckle   Telescopic Arm

Example of an extending arm manipulator

Serial Manipulator

A serial manipulator consists of a fixed base, a series of links connected by joints, and ending at a free end carrying the tool or the end-effector. In contrast to parallel manipulators, there are no closed loops. By actuating the joints, one can position and orient the end-effector in a plane or in three-dimensional (3D) space to perform desired tasks with the end-effector. This chapter deals with kinematics of serial manipulators where the motion of links are studied without considering the external forces and torques which cause these motions. The serial manipulator geometries are described using the well-known Denavit-Hartenberg (D-H) parameters. Two well-known problems, namely the direct and inverse kinematics problems, are posed, their solution procedures discussed in detail and illustrated with examples of planar and spatial serial manipulators. It is shown that closed-form analytic solutions to the inverse kinematics problem is possible only for serial manipulators with special geometries and the most general six-degree-of-freedom serial manipulator requires the solution of at most a 16th degree polynomial. The solution of the inverse kinematics problem leads to an important and useful concept of the workspace of a serial manipulator and the approaches to obtain the workspace and determine its properties are also presented.

Degrees of Freedom of a Manipulator

The degrees of freedom of a serial manipulator1 can be obtained from the well-known Chebychev- Grübler-Kutzbach criterion

$$\text{dof} = \lambda(N - J - 1) + \sum_{i=1}^{j} Fi$$

where dof is the computed degree of freedom with N as the total number of links including the fixed link (or base), J as the total number of joints connecting two consecutive links, Fi as the degrees of freedom at the ith joint, and

$\lambda = 6$, for motion in $3D$

3, for planar motion.

The quantity, dof, obtained from equation $\text{dof} = \lambda(N - J - 1) + \sum_{i=1}^{j} Fi$ is the number of independent actuators that can be present in the serial manipulator. In a broad sense, dof determines the capability of the serial manipulator with respect to dimension of the ambient space λ. We have the following possibilities:

1. dof $= \lambda -$ In this case, an end-effector of a manipulator can be positioned and oriented arbitrarily in the ambient space of motion.

2. dof $< \lambda$ – In this case, the arbitrary position and orientation of the end-effector is not achievable and there exist $(\lambda - \text{dof})$ functional relationships containing the position and orientation variables of the end-effector.

3. dof $> \lambda$ – These are called redundant manipulators and the end-effector can be positioned and oriented in infinite number of ways.

In serial manipulators with a fixed base, a free end-effector and two links connected by a joint, from equation $N = J + 1$ and dof $= \sum_{i=1}^{J} Fi$. If all the actuated joints are one-degree-offreedom joints, then $J = \text{dof}$.

If $J < \text{dof}$, then one or more of the actuated joints are multi- degree-of-freedom joints and this is not used in mechanical serial manipulators. This is due to the fact that it is difficult to locate and actuate two or more one- degree-of-freedom joints at the same place in a serial manipulator. In biological systems, muscles are used to actuate multi- degree-of-freedom joints – in a human arm muscles actuate the three- degree-of-freedom shoulder joint.

In manipulators, the J joint variables form the joint space. The variables describing the position and orientation of a link or the end-effector are called the task space variables. The dimension of task space is ≤ 6 for 3D motions and ≤ 3 for planar motion. Finally, there are often mechanical linkages, gears, etc. between actuators and joints. The space of all actuator variables is called the actuator space. If the dimension of the actuator space is more than 3 for planar motion and more than 6 for 3D motion, the manipulator is called redundant. If the dimension of the actuator space is less than the degree of freedom, then the manipulator is called under-actuated.

Servomechanism

Servomechanism refers to a device or combination of devices that automatically controls a mechanism or a source of power or energy. Servomechanisms automatically compare the controlled output of a mechanism to the controlling input. The difference between the settings or positions of the output and the input is called the error signal, which regulates the output to a desired value. Servomechanisms may be mechanical, electrical, hydraulic, or optical. The process of sending the error signal back for comparison with the input is called feedback, and the whole process of the input, output, error signal, and feedback is called a closed loop. The closed-loop system, also known as a servomechanism, has some means of incorporating mechanical feedback from the output to the input. A sensor at the output end generates a signal that is sent back to the input to regulate the machine behavior. The term servomechanism correctly applies only to systems where feedback or error-correction signals help control mechanical position or other parameters. For example, an automotive power window control is not a servomechanism, because there is no automatic feedback which controls position—the operator does this by observation. By contrast the car's cruise control uses closed loop feedback, which classifies it as a servomechanism.

Purpose of Servomechanism:

1. Accurate control of motion without the need for human attendants (automatic control),

2. Maintenance of accuracy with mechanical load variations, changes in the environment, power supply fluctuations, and aging and deterioration of components (regulation and self- calibration),

3. Control of a high-power load from a low-power command signal (power amplification) and,

4. Control of an output from a remotely located input, without the use of mechanical linkages.

A servomechanism is unique from other control systems because it controls a parameter by commanding the time-based derivative of that parameter. For example a servomechanism controlling position must be capable of changing the velocity of the system because the time-based derivative (rate change) of position is velocity. A hydraulic actuator controlled by a spool valve and a position sensor is a good example because the velocity of the actuator is proportional to the error signal of the position sensor. A simple example is the driver of a car. His eyes tell him where he is on the road, and compare it to where he should be, and this information makes its way to his brain. The brain decides what action should be taken in order to move the car from where it is to where it should be, and sends a signal to the muscles in the arm, turning the steering wheel to realign the car.

All servomechanisms have the following parts:

1. A way to measure what is desired and what is being accomplished,

2. A way to transport this information,

3. A way to determine the difference between the actual condition and the desired condition,

4. A means to amplify this difference (which is often small) and use it to move the actual condition towards the desired condition.

In the example of the car, (1) are eyes, (2) is the optic nerve and pathways to the brain, (3) is the brain, and (4) are the arms and steering wheel. A small turn of the wheel translates into a major turn for the car.

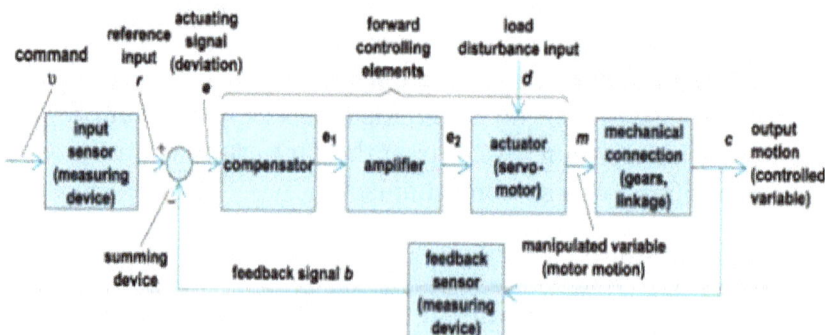

Servo loop elements and their interconnections. Cause-and-effect action takes place in the directions of arrows. (After American National Standards Institute, Terminology for Automatic Control, ANSI C85.1).

The illustration shows the basic elements of a servomechanism and their interconnections; in this type of block diagram the connection between elements is such that only a unidirectional cause-and-effect action takes place in the direction shown by the arrows. The arrows form a closed path or loop; hence this is a single-loop servomechanism or, simply, a servo loop. More complex servomechanisms may have two or more loops (multiloop servo), and a complete control system may contain many servomechanisms.

All servomechanisms have at least these basic components: a controlled device, a command device, an error detector, an error-signal amplifier, and a device to perform any necessary error corrections (the servomotor). In the controlled device, that which is being regulated is usually position. This device must, therefore, have some means of generating a signal (such as a voltage), called the feedback signal, that represents its current position. This signal is sent to an error-detecting device. The command device receives information, usually from outside the system, that represents the desired position of the controlled device. This information is converted to a form usable by the system (such as a voltage) and is fed to the same error detector as is the signal from the controlled device. The error detector compares the feedback signal (representing actual position) with the command signal (representing desired position). Any discrepancy results in an error signal that represents the correction necessary to bring the controlled device to its desired position. The error-correction signal is sent to an amplifier, and the amplified voltage is used to drive the servomotor, which repositions the controlled device. (Servomechanism may or may not use a servomotor. For example a household furnace controlled by thermostat is a servomechanism, yet there is no motor being controlled directly by the servomechanism).

In many applications, servomechanisms allow high-powered devices to be controlled by signals from devices of much lower power. The operation of the high-powered device results from a signal (called the error, or difference, signal) generated from a comparison of the desired position of the high-powered device with its actual position. The ratio between the power of the control signal and that of the device controlled can be on the order of billions to one.

Applications

Servomechanisms are useful to control motion without human attendants, or to maintain the accuracy of an environment like a power plant, and to control action from a remote isolated station. The controller typically uses (and has) much less power than that of what is being controlled. Almost always it is the position or velocity which is being controlled.

Servomechanisms are used to control mechanical things such as motors, steering mechanisms, and robots. Servomechanisms are used extensively in robotics. A robot controller can tell a servomechanism to move in certain ways that depend on the inputs from sensors. Multiple servomechanisms, when interconnected and controlled by a sophisticated computer, can do complex tasks such as cook a meal. A set of servomechanisms, including associated circuits and hardware, and intended for a specific task, constitutes a servo system. Servo systems do precise, often repetitive, mechanical chores. A computer can control a servo system made up of many servomechanisms. For example, an unmanned robotic warplane (also known as a drone) can be programmed to take off, fly a mission, return, and land. Servo systems can be programmed to do assembly-line work and other tasks that involve repetitive movement, precision, and endurance.

A servo robot is a robot whose movement is programmed into a computer. The robot follows the instructions given by the program, and carries out precise motions on that basis. Servo robots can be categorized according to the way they move. In continuous-path motion, the robot mechanism can stop anywhere along its path. In point-to-point motion, it can stop only at specific points in its path. Servo robots can be easily programmed and reprogrammed. This might be done by exchanging diskettes, by manual data entry, or by more exotic methods such as a teach box. When a robot arm must perform repetitive, precise, complex motions, the movements can be entered into the robot controller's memory. Then, when the memory is accessed, the robot arm goes through all the appropriate movements. A teach box is a device that detects and memorizes motions or processes for later recall.

Stepper Motor

A stepper motor is just another digital device, more precisely a digital DC motor. Stepper or Stepper Motor allows you to select a certain degree of movement. Rather than making a whole spin it can divide the spin into smaller parts.

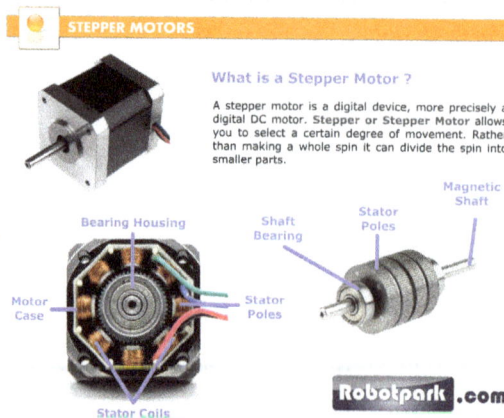

STEPPER MOTORS

What is a Stepper Motor ?

A stepper motor is a digital device, more precisely a digital DC motor. Stepper or Stepper Motor allows you to select a certain degree of movement. Rather than making a whole spin it can divide the spin into smaller parts.

Bearing Housing · Shaft Bearing · Stator Poles · Magnetic Shaft · Motor Case · Stator Poles · Stator Coils

Robotpark .com

The Stepper Motor can be commanded to hold a certain position for as a long as you want or you can just put a program to control the movements. The Stepper motor is best used in robots where only a certain degree of movement is required. One main problem with ordinary DC Motor is that these keep on moving round and round for infinity amount of time unless you stop feeding them with electricity.

If you want to build a robot that can fetch coffee for you then you can put an ordinary motor in it for moving around but for spreading his arms, collecting the coffee and delivering it safely can only be done by using a stepper motor. But if you put an ordinary motor for controlling its arms then your coffee is probably going to end up on the floor rather than on your table.

Stepper Motor Basics

Every motor converts electricity into motion, the stepper motor or step motor is named so because it converts the electricity into discrete step motion. These steps help you to choose a particular type of motion to perform.

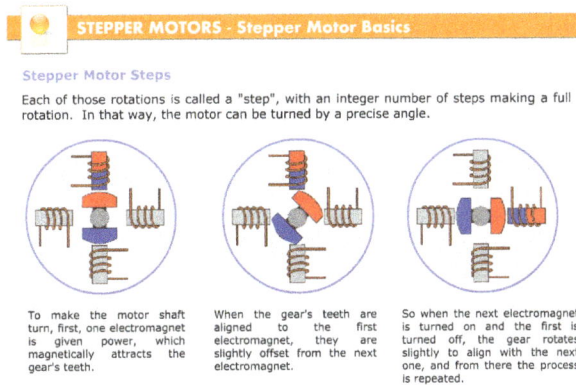

STEPPER MOTORS - Stepper Motor Basics

Stepper Motor Steps

Each of those rotations is called a "step", with an integer number of steps making a full rotation. In that way, the motor can be turned by a precise angle.

To make the motor shaft turn, first, one electromagnet is given power, which magnetically attracts the gear's teeth.

When the gear's teeth are aligned to the first electromagnet, they are slightly offset from the next electromagnet.

So when the next electromagnet is turned on and the first is turned off, the gear rotates slightly to align with the next one, and from there the process is repeated.

Stepper Motor Steps

DC Motors rotate continuously when voltage is applied to their terminals. Stepper motors, on the other hand, effectively have multiple "toothed" electromagnets arranged around a central gear-shaped piece of iron. The electromagnets are energized by an external control circuit, such as a microcontroller.

The Stepper Motor itself is controlled by a driver which sends the electric pulses to the motor that shows movement in response to these electric pulses. The pulse frequency decides the total amount of movement and rotation. The frequency is controlled by the driver which in turn is controlled by a human that means you are controlling the movement directly.

To make the motor shaft turn, first, one electromagnet is given power, which magnetically attracts the gear's teeth. When the gear's teeth are aligned to the first electromagnet, they are slightly offset from the next electromagnet. So when the next electromagnet is turned on and the first is turned off, the gear rotates slightly to align with the next one, and from there the process is repeated. Each of those rotations is called a "step", with an integer number of steps making a full rotation. In that way, the motor can be turned by a precise angle.

Stepper motors can respond and accelerate quickly. They have low rotor inertia that can get up to speed quickly. For this reason step motors are ideal for short, quick moves.

One Step of a Stepper Motor

A stepper motor (or step motor) is a brushless DC electric motor that divides a full rotation into a number of equal steps. Every turn of the motor is divided into a discrete number of steps, in many cases 200 steps. The stepper motor driver sends a special pulse to the Stepper Motor for each step. Since each pulse causes the motor to rotate a precise angle, typically 1.8°, the motor's position can be controlled without any feedback mechanism. As the digital pulses increase in frequency, the step movement changes into continuous rotation, with the speed of rotation directly proportional to the frequency of the pulses.

Stepper Motor Advantages

- It is very safe to use.
- It is very easy to setup.
- It provides with highly controlled movement.
- The Stepper motors have very long life, only if you don't break it by an accident.
- It is an excellent repeater device it can repeat its movements very accurately.
- Stepper motor is also very cheap as compared to other motion controlling devices.
- One of the main features of stepper motor is that it can't be damaged by overloading.
- When overloaded it stops working until the load is removed.
- It is easy to operate.
- It offers very precise movement.

Stepper Motor Disadvantages

- Resonance occurs if not properly controlled.
- Have low efficiency; it draws more power than the out-put it provides.

- Its torque is inversely proportional to its speed.

- It can't accelerate load rapidly.

- It becomes very difficult to operate at high speeds.

Stepper Motor Accuracy

One of the most remarkable features of the step motor is its precision, the stepper motors are very accurate. But like all other human-made things these have some errors. The precision is not totally awe-inspiring it has chances of error. Standard step motors have an accuracy of ± 3 arc minutes (0.05°).

The remarkable feature of steps motors, though, is that this error does not accumulate from step to step. When a standard step motor travels one step, it will go 1.8° ± 0.05°. If the same motor travels one million steps, it will travel 1,800,000° ± 0.05°. The error does not accumulate. Mostly the error percentage is about pulse minus 0.005. These error chances remain same whether motor takes ten steps or million.

Stepper Motor Applications

Step motors are used every day in both industrial and commercial applications because of their low cost, high reliability, high torque at low speeds. For making more advanced technologies we use stepper motors rather than simple motors. Computer-controlled stepper motors are a type of motion-control positioning system. Step Motors are typically digitally controlled motors, for use in holding or precise positioning applications. These motors are used at:

-Industrial Machines: CNC Machines, Milling Machines, Laser Cutters etc.

-Computer Technology: CD-Roms, DVD Players, Floppy Disk Drives, Scanners etc.

-Printing: Printers, Plotters, 3D Printers etc.

-Intelligent Lighting Systems: Lasers, Optical Devices, Mirror Mounts.

Locomotion in Robots

Locomotion involves the conversion of some source of energy — electricity, air pressure, steam, nuclear power. — into a mechanical action that moves a vehicle or other carriage. Consider the lowly car: gas you put into the tank is converted to mechanical power by means of internal combustion. The gas is compressed as a vapor and explodes against a cylinder. The explosion pushes the cylinder down; this cylinder is in turn is connected to a drive shaft, which spins the wheels. The process repeats itself thousands of times per minute.

Mobile robots use a variety of techniques to achieve motion. Most use an electric power source (usually a battery) that operates an electric motor. In the typical arrangement the direction of the motors can be changed, allowing the robot to be propelled forward or backward. There are other power train techniques used for robots, but the battery and motor pair is by far the most common.

Wheels

Wheels are the most popular method of providing robot mobility. Robot wheels can be just about any size, dictated only by the dimensions of the robot and your outlandish imagination. However, for reasons of practicality and weight, small robots usually have small wheels, less than two or three inches in diameter. Medium sized robots use wheels with diameters up to seven or eight inches. A few unusual designs call for bicycle wheels, which despite their size are lightweight but very sturdy.

Robots can have just about any number of wheels, although two is the most common. The robot is balanced on its wheels by one or two skids or swivel casters. Four and six wheel robots are also around. The most common robot uses two drive wheels. Placed on either side.

Legs

A small percentage of amateur robots are designed with legs, and such robots can be conversation pieces all of their own. Many difficulties must be overcome in designing and constructing a legged robot. First, there is the question of the number of legs, and how the legs provide stability when the robot is in motion. Or more critical, stability when it's standing still. Then there is the question of how the legs propel the robot forward or backward — and more difficult still! — The question of how to turn the robot so it can navigate a corner.

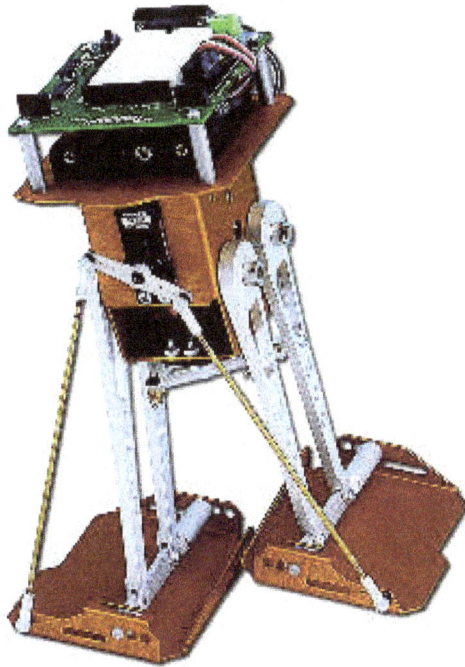

Two-legged walking robot

Legged robots are a challenge to design and build, but they provide an extra level of mobility that wheeled robots cannot. Wheel-based robots may have a difficult time navigating through rough terrain, but a properly designed leg-based robot can easily walk right over small ditches and obstacles.

A few daring robot experimenters have come out with two-legged robots, but the difficulties in assuring balance and control make these designs largely impractical for most robot hobbyists. Exception: the robot uses wide "duck" feet, and waddles as it walks. With this arrangement, however, turning can be inexact.

Four-legged robots (quadrupeds) are easier to balance, but good locomotion and steering can be difficult to achieve. I've found robots with six legs (called hexapods) are able to walk at brisk speeds without falling, and are more than capable of turning corners, bounding over uneven terrain, and making the neighborhood dogs and cats hide for cover.

Robots with legs add many complexities.

Tracks

The basic design of track driven robots is simple: two tracks, one on each side of the robot, act as giant wheels. The tracks turn, and the robot lurches forward or backward. For maximum traction, each track is the same length, or somewhat shorter, than the length of the robot itself — though many variations are possible.

Tankbot Servo from Budget Robotics

Track drive is practical for many reasons, including the ability to mow through all sorts of obstacles, like rocks, ditches, and potholes. Given the right track material, traction is excellent, even on slippery surfaces like snow, wet concrete, or a clean kitchen floor.

For the most part, constructing and effective track drive is harder than implementing wheels. The reason: the tracks present a large contact area. This larger contact area increases traction when moving forward or backward, but it also restricts turning. Tracked vehicles, like tanks, turn by skidding or slipping around a turning point — hence they are referred to as having skid-steering. If the treads are super-pliable, and the surface is hard (like a kitchen floor), the added friction can greatly impair the ability of the vehicle to turn.

Zero Moment Point

Zmp for Bipedal Robots

The Zero-Moment Point, or ZMP, is the point on the surface of the foot where a resultant force R can replace the force distribution shown in figure below. Mathematically, the ZMP can be calculated from a group of contact points p_i for $i = 1, \ldots, N$ with each force vector f_i associated with the contact point,

$$ZMP = \frac{\sum_{i=1}^{N} p_i f_{iz}}{\sum_{i=1}^{N} f_{iz}} = \frac{\sum_{i=1}^{N} p_i f_{iz}}{f_z}.$$

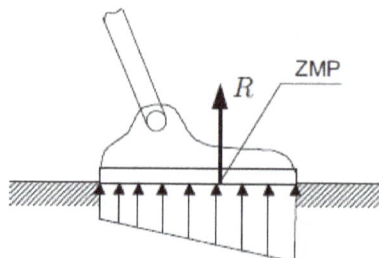

Figure. Definition of Zero-Moment Point

According to this, the ZMP can never leave the support polygon. If the floor is assumed horizontal, the torque reduces to

$$\tau_x = \tau_y = 0$$

at the ZMP. Further derivations for the torque at the ZMP can be found in Appendix A.1. This definition is useful when pressure sensors are attached to the feet. With these sensors, the center of pressure can be calculated on the feet, and the ZMP can be directly measured. The humanoid robot DARwIn-OP has pressure sensor foot attachments that can be bought separately. However, these sensors were not used, so a model-based method was employed to calculate the ZMP.

Computed ZMP

Instead of calculating the resultant force from the force distribution, the position of the ZMP can be calculated from the dynamics of the system using Newton's Second Law when the mass properties and motion of the robot are known. To illustrate the use of computed ZMP, the cart-table model used in the controller is presented in figure. This model includes a cart of mass M that rolls without slipping along a flat table with a height of z_c. The table does not rotate as long as point p, denoting the ZMP, stays inside the base, inferring that $\tau_{zmp} = 0$. If the horizontal position of the center of mass is given by x relative to origin O, the net torque around point p can be calculated from the vector equation

$$\tau_p = p \times R + \tau_{zmp}.$$

This expression is simplified to the scalar equation

$$\tau_{zmp} = -Mg\,(x - p) + Mz_c\ddot{x}$$

Using Eq. $\tau_x = \tau_y = 0$, Eq. $\tau_{zmp} = -Mg\,(x - p) + Mz_c\ddot{x}$ can be simplified and solved to give an expression for the ZMP,

$$p = x - \frac{z_c}{g}\,\ddot{x}.$$

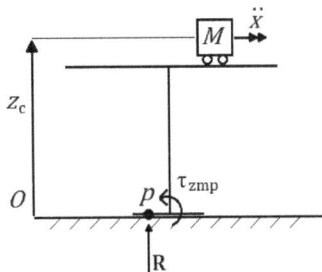

Figure: Cart-Table Model

Looking closer at Eq. $\tau_{zmp} = -Mg(x - p) + Mz_c\ddot{x}$, two interesting observations emerge about the computed ZMP from this definition.

1. The ZMP becomes the CoM when there is no acceleration.

2. The ZMP can be located outside the support polygon.

If the ZMP leaves the support polygon during the gait, the motion is determined to be unstable because it can cause the robot to tip over about an edge of the support polygon. This, however, does not necessarily mean that the robot will fall. For example, if a bipedal robot has feet glued to the ground, the ZMP could be outside the support polygon without the robot falling. The cohesive force of the glue adds an extra force to balance the torque around the ZMP so that the robot does not fall until the glue fails. If falling, the robot can recover stable walking either through its motion (compensating body movement) or increasing its support polygon (placing the other foot on the ground).

The ZMP can also be calculated for the 3D case using Newton's and Euler's laws for the change in linear and angular momentum. More general expressions, similar to Eq. 1.5, were presented in and summarized in Appendix A.2. This was used in modeling the full 3D dynamics of DARwIn-OP to calculate the ZMP. However, this 3D model was never formally used in the controller, but was developed using Kane's method.

The Human Gait and a Simplified Model

One of the goals of implementing this humanoid walking controller is to achieve a walking gait that is more human-like. In order to achieve this, the human gait is presented and then simplified to match the needs of the controller. During walking, at least one foot remains in contact with the ground at all times. Each step consists of two main phases, stance and swing, consisting of about 62% and 38% of the gait cycle, respectively. Each of these phases can be subdivided into 8 more categories : initial contact, loading response, midstance, terminal stance, preswing, initial swing, midswing, and terminal swing as shown in figure.

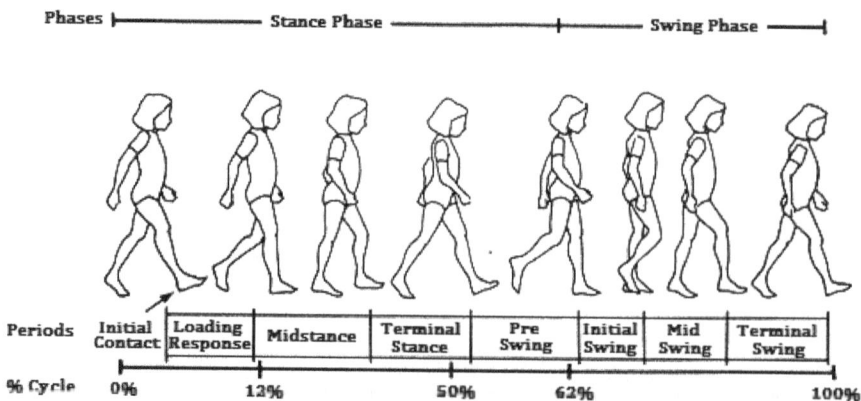

Figure: Human Gait Cycle

When the leading foot touches the ground, initial contact begins, starting the stance phase. Usually, this involves a simple heel strike to the ground. The main implusive force from ground contact is experienced in the next phase, the loading response. In this phase, the leading foot flattens, coming in full contact with the ground. The leading foot then absorbs the contact force from the ground. After this, the first single support phase, called midstance, begins when the trailing foot leaves the ground and ends when the body weight is aligned over the front of the current stance foot. During midstance, the body has a maximum potential energy as the body's center of gravity reaches its maximum height, rising over the stance leg. The terminal stance phase begins as soon as the body's weight shifts past the current stance foot. Heel off occurs during this phase when the stance foot's heel leaves the ground. The final stance phase is preswing, in which the previous stance foot prepares to lift off, but is still in contact with the ground. This is called toe off as the previous stance foot's toe leaves the ground and can no longer provide active forward propulsion.

Now, the previous stance leg during single support becomes the swing foot for the next three phases. Initial swing begins immediately after the swing foot leaves the ground and lasts until maximum knee flexion, where midswing starts. During midswing, the swing foot is brought past the stance foot, making sure that the foot has enough ground clearance. The phase ends when the tibia is perpendicular to the ground. The final phase, terminal swing, prepares the body for contact with the ground and fully extends the knee. This completes one gait cycle.

In order to simplify the human gait cycle, the proposed controller is broken into four distinct phases: right leg double support, right leg single support, left leg double support, and left leg single support. Relating these to the aforementioned phases, right leg double support tries to capture initial contact to midstance. Right leg single support refers to the period of time during which the left leg is swinging through the air, namely midstance and most of terminal stance. As soon as the left swing leg touches the ground, left leg double support begins, which includes the last part of terminal stance and preswing. The final part of the gait, left leg single support, includes the entire swing phase from initial swing to terminal swing. The controller develops schemes to handle these four phases throughout the walking gait cycle.

Robotic Non-destructive Testing

Non-destructive testing (NDT) is a highly multidisciplinary group of analysis techniques used throughout science and industry to evaluate the properties of materials, and/or to ensure the integrity of components/structures, without causing damage to them. In civil aerospace manufacturing, the increasing deployment of composite materials demands a high integrity and traceability of NDT measurements, combined with a rapid throughput of data. Modern components increasingly present challenging shapes and geometries for inspection. Using traditional manual inspection approaches produce a time-consuming bottleneck in the industrial production and this limitation provides the fundamental motivation for increased automation.

Modern Computer-Aided Design (CAD) is used extensively in composite manufacture. Additionally, where it was once necessary to construct large items from many smaller parts, Computer-Aided Manufacturing (CAM) now allows these large items to be produced easily from one piece of raw material (through traditional subtractive approaches, or built up using more recent additive manufacturing processes). As a result, large components with complex geometries are becoming very common in modern structures, and the aerospace industry is a typical field, where wide complex shaped parts are very frequently used. Moreover the use of composite materials, which are notoriously challenging to inspect, is becoming widespread in the construction of new generations of civilian aircraft. To cope with future demand projections for these operations, it is therefore essential to overcome the current NDT bottleneck, which traditionally can be the slowest aspect in a production process.

A fundamental issue with composites manufacturing compared to conventional light alloy materials lies in the process variability. Often parts that are designed as identical, will have significant deviations from CAD, and also may change shape when removed from the mould. This presents a significant challenge for precision NDT measurement deployment which must be flexible to accommodate these manufacturing issues.

For these reasons, NDT inspection is often performed manually by technicians who typically have to position and move appropriate probes over the contour of the sample surfaces. Manual scanning requires trained technicians and results in a very slow inspection process for large samples. The repeatability of a test can be challenging in structures where complex setups are necessary to perform the inspection (e.g. orientation of the probe, constant standoff, etc.). While manual scanning may remain a valid approach around the edges of a structure, or the edges of holes in a structure, developing reliable automated solutions has become an industry priority to drive down inspection times. The fundamental aims of automation within the inspection process are to minimize downtimes due to the higher achievable speed, and to minimize variability due to human factors.

Semi-automated inspection systems have been developed to overcome some of the shortcomings with manual inspection techniques, using both mobile and fixed robotic platforms. The use of linear manipulators and bridge designs has, for a number of years, provided the most stable conditions in terms of positioning accuracy. The use of these systems to inspect parts with noncomplex shapes (plates, cylinders or cones) is widespread; typically, they are specific machines, which are used to inspect identically shaped and/or sized parts.

More recently, many manufacturers of industrial robots have produced robotic manipulators with excellent positional accuracy and repeatability. An industrial robot is defined as an automatically controlled, reprogrammable, multipurpose manipulator, programmable in three or more axes . In the spectrum of robot manipulators, some modern robots have suitable attributes to develop automated NDT systems and cope with the challenging situations seen in the aerospace industry. They present precise mechanical systems, the possibility to accurately master each joint, and the ability to export positional data at frequencies up to 500 Hz. Some applications of 6-axis robotic arms in the NDT field have been published during the last few years and there is a growing interest in using such automation solutions with many manufacturers within the aerospace sector.

Exploring the current state of the art, RABIT is a group of systems developed by TECNATOM S.A., in collaboration with KUKA Robots Ibérica, that first approached the possibility of incorporating the use of industrial robots in NDT applications . These systems boast the capability of using the potential of industrial robots and integrating them in an overall inspection apparatus, bringing together all the hardware and software required to plan and configure ultrasonic inspections. Off-the-shelf robotic arms were also used in the Laser Ultrasound for Composite InspEction (LUCIE) system, addressed to inspect large curved surfaces such as the inside of aircraft fuselage, by means of ultrasound generated by laser. Genesis Systems Group has developed the NSpect family of Robotic Non-Destructive Inspection cells. Incorporating the FlawInspecta technology, developed by Diagnostic Sonar in conjunction with National Instruments, the NSpect systems employ a KUKA 6DOF robot arm to perform ultrasonic inspection using either an immersion tank, or a recirculating water couplant. General Electric (GE) has also investigated the integration of phased array UT with off-the-shelf industrial robots for the inspection of aerospace composites.

Despite these previous efforts, there remain challenges to be addressed before fully automated NDT inspection of complex geometry composite parts becomes commonplace. The key challenges include generation and in-process modification of the robot tool-path, high speed NDT data collection, integration of surface metrology measurements, and overall visualization of measurement results in a user friendly fashion. Collaborations driving this vision include TWI Technology Centre (Wales), which is currently carrying out a 3-year project, called IntACom, on behalf of its sponsors; its objective is to achieve a fourfold increase in the throughput of aerospace components. Additionally the UK RCNDE consortium conducts research into integration of metrology with NDT inspection. Both these consortia have identified the requirement for optimal tool path generation over complex curved surfaces.

Approaches to Robotic Path Planning for NDT Applications

Existing Robotic Path Planning Software

Six-axis robotic arms have traditionally been used in production lines to move the robot

end-effector from one position to another for repetitive assembly and welding operations. In this scenario, where the exact trajectory between two points in the space is not
too important, the teach pendant of a robot is used to move manually the end-effector to
the desired position and orientation at each stage of the robot task. Relevant robot configurations are recorded by the robot controller and a robot programme is then written to
command the robot to move through the recorded end-effector postures. More recently,
accurate mechanical joints and control units have made industrial robotic arms flexible
and precise enough for finishing tasks in manufacturing operations. Robotic manipulators are highly complex systems and the trajectory accuracy of a machining tool has a
huge impact on the quality and tolerances of the finished surfaces. As a result, many software environments have been developed by manufacturers, academic researchers and
also by the robot manufacturers themselves, in order to help technicians and engineers
to programme complex robot tasks. The use of such software platforms to programme
robot movements is known as off-line programming (OLP). It is based on the 3D virtual
representation of the complete robot work cell, the robot end-effector and the samples to
be manipulated or machined. Although some limited applications for inspection delivery
have been demonstrated, in general conventional OLP is geared towards manufacturing
applications where the task is the production of a specific component using conventional
milling/ drilling / trimming operations. In contrast, the result of an automated NDT inspection requires a flexible and extensible approach that has the flexibility to allow future
changes in the path planning to accommodate requirements of future NDT inspections.
Building the NDT toolbox in MATLAB provides an easy route to such future adaptation
by the research community (e.g. fluid dynamics modelling of water-jets, compensation
for part variability and conditional programming approaches).

Using current OLP software to generate appropriate tool-paths for NDT purposes can
appear quite straightforward at first inspection; however it is possible to list a series of
serious inadequacies:

1. Path-planning for automated NDT inspections is a very specific task. As previously mentioned, much commercial software for off-line robot programming
   draws its origin from the need to use the advantageous flexibility of general
   robotic manipulators to replace the more traditional and usual machining tools
   (milling machines, lathes, etc.). As a result, many commercial software applications for off-line programming of robots are expensive tools, incorporating
   a lot of functionality specific for CAD/CAM purposes and machining features.
   Conventional OLP software has no easy provision for modification of tool path
   based on specific characteristics of NDT inspection. For example, in water-jet
   coupled ultrasonic testing, the exact orientation and separation from part of the
   end effector, combined with water pressure and nozzle characteristics will influence the trajectory of the water jet. Since the new toolbox is MATLAB based,
   it is flexible and easy to accommodate such new features into the path planning
   algorithm in future.

2. Significant complications exist when two or more robotic arms need to be synchronized in order to perform a specific NDT inspection. The Ultrasonic Through-Transmission (UTT) technique, for example, uses two transducers: one emitter and one receiver; the receiver being placed on the opposite side of the component and facing the transmitting probe. Many commercial pieces of software (e.g. Delcam and Mastercam) do not offer any support for co-operating robots. Fast Surf, an add-on from CENIT for Delmia (Dassault Systems), allows partial synchronization of robotic movements (e.g. at start or end points of complex paths) but not full synchronization over the complete path, required for the UTT technique. Our new MATLAB toolbox provides the capability for full point to point synchronization between robots for situations where a change in material section is encountered – this is far more sophisticated than simple master–slave synchronization implemented by typical robot equipment suppliers.

Robotic Sensors

Robotic sensors are used to estimate the robot's condition and environment, These signals are passed to the controller to enable the appropriate behavior, The sensors in the robots are based on the functions of the human sensory organs, The robots require the extensive information about their environment in order to function effectively.

Robots need Robot Sensors to know the world around them, There are many Robot Sensors including ultrasonic, the temperature & the humidity, the force and a lot more to increase the robot awareness.

Types of Robot Sensors

Light Sensors

A Light sensor is used to detect light and create a voltage difference. The two main light sensors generally used in robots are Photoresistor and Photovoltaic cells. Other kinds of light sensors like Phototubes, Phototransistors, CCD's etc. are rarely used.

Photoresistor is a type of resistor whose resistance varies with change in light intensity; more light leads to less resistance and less light leads to more resistance. These inexpensive sensors can be easily implemented in most light dependant robots.

Photovoltaic cells convert solar radiation into electrical energy. This is especially helpful if you are planning to build a solar robot. Although photovoltaic cell is considered as an energy source, an intelligent implementation combined with transistors and capacitors can convert this into a sensor.

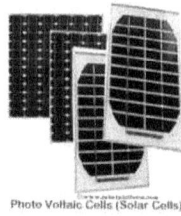

Photo Voltaic Cells (Solar Cells)   Photoresistors (LDR)

Sound Sensor

As the name suggests, this sensor (generally a microphone) detects sound and returns a voltage proportional to the sound level. Sound Sensor A simple robot can be designed to navigate based on the sound it receives. Imagine a robot which turns right for one clap and turns left for two claps. Complex robots can use the same microphone for speech and voice recognition.

www.robotplatform.com
Sound Sensor (Mic)

Implementing sound sensors is not as easy as light sensors because Sound sensors generate a very small voltage difference which should be amplified to generate measurable voltage change.

Temperature Sensor Temperature Sensors

What if your robot has to work in a desert and transmit ambient temperature? Simple solution is to use a temperature sensor. Tiny temperature sensor ICs provide voltage difference for a change in temperature. Few generally used temperature sensor IC's are LM34, LM35, TMP35, TMP36, and TMP37.

www.robotplatform.com
Temperature Sensors

Contact Sensor

Contact sensors are those which require physical contact against other objects to trigger. A push button switch, limit switch or tactile bumper switch are all examples of contact sensors. Limit Switch These sensors are mostly used for obstacle avoidance robots. When these switches hit an obstacle, it triggers the robot to do a task, which can be reversing, turning, switching on a LED, Stopping etc. There are also capacitive contact sensors which react only to human touch (Not sure if they react to animals touch). Touch screen Smart phones available these days use capacitive touch sensors. Contact Sensors can be easily implemented, but the drawback is that they require physical contact. In other words, your robot will not turn until it hits an object. A better alternative is to use a proximity sensor.

© www.robotplatform.com
Limit Switch (Contact Sensor)

Proximity Sensor

This is a type of sensor which can detect the presence of a nearby object within a given distance, without any physical contact. The working principle of a Proximity sensor is simple. A transmitter transmits an electromagnetic radiation or creates an electrostatic field and a receiver receives and analyzes the return signal for interruptions. There are different types of Proximity sensors and we will discuss only a few of them which are generally used in robots.

- Infrared (IR) Transceivers: An IR LED transmits a beam of IR light and if it finds an obstacle, the light is simply reflected back which is captured by an IR receiver. Few IR transceivers can also be used for distance measurement.

- Ultrasonic Sensor: These sensors generate high frequency sound waves; the received echo suggests an object interruption. Ultrasonic Sensors can also be used for distance measurement.

- Photoresistor: Photoresistor is a light sensor; but, it can still be used as a proximity sensor. When an object comes in close proximity to the sensor, the amount of light changes which in turn changes the resistance of the Photoresistor. This change can be detected and processed.

There are many different kinds of proximity sensors and only a few of them are generally preferred for robots. For example, Capacitive Proximity sensors are available which detects change in capacitance around it. Inductive proximity sensor detects objects and distance through the use of induced magnetic field.

Distance Sensor

Most proximity sensors can also be used as distance sensors, or commonly known as Range Sensors; IR transceivers and Ultrasonic Sensors are best suited for distance measurement.

- Ultrasonic Distance Sensors: The sensor emits an ultrasonic pulse and is captured by a receiver. Since the speed of sound is almost constant in air, which is 344m/s, the time between send and receive is calculated to give the distance between your robot and the obstacle. Ultrasonic distance sensors are especially useful for underwater robots.

- Infrared Distance sensor: IR circuits are designed on triangulation principle for distance measurement. A transmitter sends a pulse of IR signals which is detected by the receiver if there is an obstacle and based on the angle the signal is received, distance is calculated. SHARP has a family of IR transceivers which are very useful for distance measurement. A simple transmit and receive using a couple of transmitters and receivers will still do the job of distance measurement, but if you require precision, then prefer the triangulation method.

- Laser range Sensor: Laser light is transmitted and the reflected light is captured and analyzed. Distance is measured by calculating the speed of light and time taken for the light to reflect back to the receiver. These sensors are very useful for longer distances.

- Encoders: These sensors (not actually sensors, but a combination of different components) convert angular position of a shaft or wheel into an analog or digital code. The most popular encoder is an optical encoder which includes a rotational disk, light source and a light detector (generally an IR transmitter and IR receiver). The rotational disk has transparent and opaque pattern (or just black and white pattern) painted or printed over it. When the disk rotates along with the wheel the emitted light is interrupted generating a signal output. The number of times the interruption happens and the diameter of the wheel can together give the distance travelled by the robot.

- Stereo Camera: Two cameras placed adjacent to each other can provide depth information using its stereo vision. Processing the data received from a camera is difficult for a robot with minimal processing power and memory. If opted for, they make a valuable addition to your robot.

There are other stretch and bend sensors which are also capable of measuring distance. But, their range is so limited that they are almost useless for mobile robots.

Pressure Sensors

As the name suggests, pressure sensor measures pressure. Tactile pressure sensors are useful in robotics as they are sensitive to touch, force and pressure. If you design a robot hand and need to measure the amount of grip and pressure required to hold an object, then this is what you would want to use.

Tilt Sensors

Tilt sensors measure tilt of an object. In a typical analog tilt sensor, a small amount of mercury is suspended in a glass bulb. When mercury flows towards one end, it closes a switch which suggests a tilt.

Navigation/Positioning Sensors

The name says it all. Positioning sensors are used to approximate the position of a robot, some for indoor positioning and few others for outdoor positioning.

- GPS (Global Positioning System): The most commonly used positioning sensor is a GPS. Satellites orbiting our earth transmit signals and a receiver on a robot acquires these signals and processes it. The processed information can be used to determine the approximate position and velocity of a robot. These GPS systems are extremely helpful for outdoor robots, but fail indoors. They are also bit expensive at the moment and if their prices fall, very soon you would see most robots with a GPS module attached.

- Digital Magnetic Compass: Similar to a handheld magnetic compass, Digital Magnetic compass provides directional measurements using the earth's magnetic field which guides your robot in the right direction to reach its destination. These sensors are cheap compared to GPS modules, but a compass works best along with a GPS module if you require both positional feedback and navigation. Philips KMZ51 is sensitive enough to detect earth's magnetic field.

- Localization: Localization refers to the task of automatically determining the location of a robot in complex environment. Localization is based on external elements called landmarks which can be either artificially placed landmarks, or natural landmark. In the first approach, artificial landmarks or beacons are placed around the robot, and a robot's sensor captures these signals to determine its exact location. Natural landmarks can be doors, windows, walls, etc. which are sensed by a robots sensor / vision system (Camera). Localization can be achieved using beacons which generate Wi-Fi, Bluetooth, Ultrasound, Infrared, Radio transmissions, Visible Light, or any similar signal.

Acceleration Sensor

An accelerometer is a device which measures acceleration and tilt. There are two kinds of forces which can affect an accelerometer: Static force and Dynamic Force.

Static Force: Static force is the frictional force between any two objects. For example earth's gravitational force is static which pulls an object towards it. Measuring this gravitational force can tell you how much your robot is tilting. This measurement is exceptionally useful in a balancing robot, or to tell you if your robot is driving uphill or on a flat surface.

Dynamic force: Dynamic force is the amount of acceleration required to move an object. Measuring this dynamic force using an accelerometer tells you the velocity/speed at which your robot is moving. We can also measure vibration of a robot using an accelerometer, if in any case you need to.

Accelerometer comes in different flavors. Always select the one which is most appropriate for your robot. Some of the factors which you need to consider before selecting an accelerometer are:

1. Output Type: Analog or Digital

2. Number of Axis: 1,2 or 3

3. Accelerometer Swing: ±1.5g, ±2g, ±4g, ±8g, ±16g

4. Sensitivity: Higher or Lower (Higher the better)

5. Bandwidth

Gyroscope

A gyroscope or simply Gyro is a device which measures and helps maintain orientation using the principle of angular momentum. In other words, a Gyro is used to measure the rate of rotation around a particular axis. Gyroscope is especially useful when you want your robot to not depend on earth's gravity for maintaining Orientation. (Unlike accelerometer).

IMU

Inertial Measurement Units combine properties of two or more sensors such as Accelerometer, Gyro, Magnetometer, etc., to measure orientation, velocity and gravitational forces. In simple words, IMU's are capable of providing feedback by detecting changes in an objects orientation (pitch, roll and yaw), velocity and gravitational forces. Few IMUs go a step further and combine a GPS device providing positional feedback.

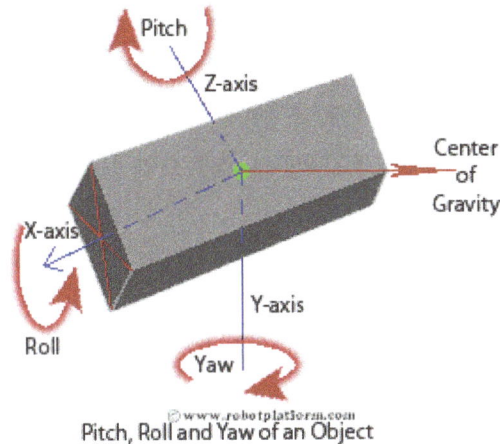

Pitch, Roll and Yaw of an Object

Voltage Sensors

Voltage sensors typically convert lower voltages to higher voltages, or vice versa. One example is a general Operational-Amplifier (Op-Amp) which accepts a low voltage, amplifies it, and generates a higher voltage output. Few voltage sensors are used to find the potential difference between two ends (Voltage Comparator). Even a simple LED can act as a voltage sensor which can detect a voltage difference and light up. (not considering current requirements here).

Current Sensors

Current sensors are electronic circuits which monitor the current flow in a circuit and output either a proportional voltage or a current. Most current sensors output an analog voltage between 0V to 5V which can be processed further using a microcontroller.

Other Sensors for Robots

There are hundreds of sensors made today to sense virtually anything you can think of, and it is almost impossible to list all available sensors. Apart from those mentioned above, there are many other sensors used for specific applications. For example: Humidity Sensors measures Humidity; Gas sensors are designed to detect particular gases (helpful for robots which detects gas leaks); Potentiometers are so versatile that they can be used in numerous different applications; Magnetic Field Sensors detect the strength of magnetic field around it.

References

- Types-of-robotic-actuators-8253: roboticstomorrow.com, Retrieved 12 May 2018

- What-do-delta-robots-have-in-common-with-spiders: linearmotiontips.com, Retrieved 28 April 2018

- Tools-as-end-effectors: roboticsbible.com, Retrieved 17 July 2018

- Stepper-Motors: robotpark.com, Retrieved 14 June 2018

- Exploring-robot-locomotion-systems: robotoid.com, Retrieved 24 May 2018

- Types-of-robot-sensors: robotplatform.com, Retrieved 10 April 2018

4

Understanding Mechatronics

Mechatronics is a field of engineering that unifies the principles and methods of electronics, mechanics, robotics and computing to develop more economical and simpler systems. An important example of a mechatronic system is an industrial robot. This chapter explores the fundamentals of mechatronics, modeling of mechatronic systems and applications of mechatronics.

Mechatronics refers to the successful combination of mechanical systems and electronics. In Mechatronics, traditional systems of mechanical engineering are fused together with components from computer science, mathematics and electrical engineering. Mechatronics has a bright future and is currently applied in everyday life for solutions ranging from transportation to optical telecommunication and biomedical engineering.

A Representational Image of Mechatronics System

Mechatronics provides solutions that are efficient and reliable systems. Mechatronic Systems mostly have microcomputers to ensure smooth functioning and higher dependability. The sensors in these systems absorb signals from the surroundings, react to these signals using appropriate processing to generate acquired output signals. Few examples of Mechatronics System are automated guided vehicles, robots, digitally controlled combust engines and machine tools with self-adaptive tools, aircraft flight control and navigation systems, and smart home appliances (e.g. Washers, dryers, etc.).

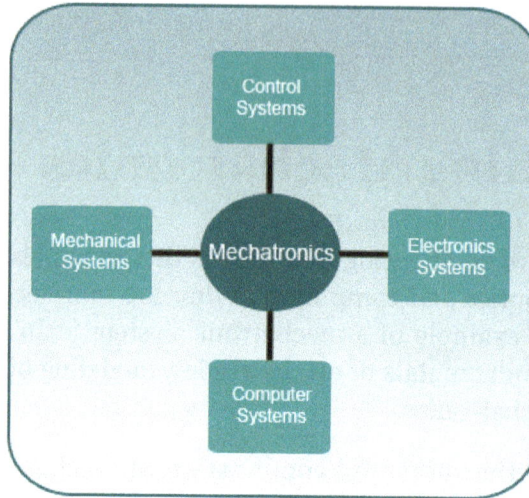

An Overview of Various Fields that Makes Mechatronics

The control system is part of the loop for operating continously in the surroundings. The mechanical systems contributes design, manufacturing and system dynamics. Computers contribute data acquisition method and algorithms. Electrical systems include DC and AC circuit analysis, power analysis and semiconductor device analysis.

The requirements for the life cycle of mechatronics design includes:

- Upgradeability

- Reliability

- Service protocols and methods which include on board diagnostics, prognos

- Maintenance

- Delivery parameters such as time, cost and design

- Disposal processes which includes recycling, and disposing hazardous materials

Importance of Mechatronics

As mentioned above, mechatronics is an integration of electronics, computer science and mechanical engineering. The manufacturing industries totally depend on the integration of computer and electronics technologies for better products and processes. As the situation became very competitive it was necessary to segregate the electronics and mechanical branches. This division of the two branches led to an interdisciplinary approach to introduce mechatronics. Mechatronics can be referred as Elctromechanical systems or control automotive engineering.

Working

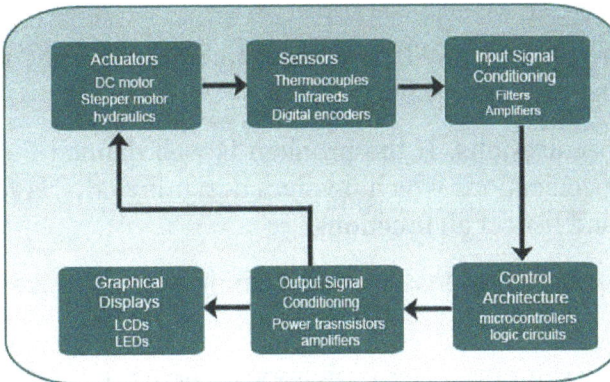

Simple Diagram Showing Diffrent Functions of Mechatronis System

The mechatronics system includes:

- Measurement and actutation module: Signals are received from the external world and feedback signal. This segment consists of actuators and sensors like stepper motors, solenoids, AC/DC, strain gauge, temperature sensor/pressure sensors/photo sensors.

- Communication Module: The relative position of actuators and position of sensors can be measured to generate appropriate signals. These signals are transferred through the communication module to the CPU. The signal conditioning circuits, interfacing circuits and bus communication form the communication module.

- CPU: When the CPU receives the signal, they can perform logical and arithmetic operations with a processor and software. An appropriate control signal is generated by the CPU.

- Output signal conditioning module: This module includes amplifiers to drive plotter, audio-visual indicators, ADC/DAC's and displays. The output signal is forwarded to feedback module.

- Feedback module: This module generates signal proportional to the output signal which is forwarded to the measurement and actuation module. The signal from the external environment and the feedback signal are compared by the measurement and actuation module.

System Design

Mechatronics System Design

The design process of mechatronics system involves a number of stages. The important stages are as follows:

- Identify the need: The first step while designing a product is to identify the need of the system. Market survey or market research can be done to recognize the need.

- Analyzing the problem: With the available data from the market survey, the problems can be analyzed appropriately.

- Preparing specifications: If the problem is well defined then the next step is to prepare specifications which involves designing criteria for quality, existing constraints and lists of all functions.

- Generate possible solutions: In the designing process, this stage is the theoretical one.

- Select suitable solution: After evaluations, one appropriate solution has to be chosen depending on the criteria.

- Prepare detailed design and drawing: The design and drawing of the selected alternative is prepared.

- Implement the design: The chosen design can be transformed into working drawing and practical circuits to produce the product.

Biomechatronics

Biomechatronics is the interdisciplinary study of biology, mechanics, electronics and control. It focuses on the research and design of assistive, therapeutic and diagnostic devices to compensate (partially) for the loss of human physiological functions or to enhance these functions.

Biomechatronics is the merging of man with machine -- like the cyborg of science fiction. It is an interdisciplinary field encompassing biology, neurosciences, mechanics, electronics and robotics. Biomechatronic scientists attempt to make devices that interact with human muscle, skeleton, and nervous systems with the goals of assisting or enhancing human motor control that can be lost or impaired by trauma, disease or birth defects.

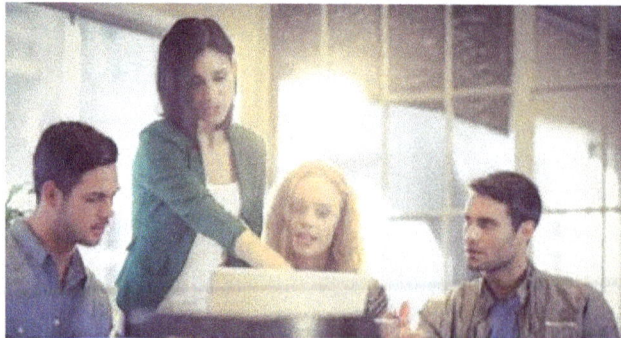

Consider what happens when you lift your foot to walk:

- The motor center of your brain sends impulses to the muscles in your foot and leg. The appropriate muscles contract in the appropriate sequence to move and lift your foot.

- Nerve cells in your foot sense the ground and feedback information to your brain to adjust the force, or the number of muscle groups required to walk across the surface. You don't apply the same force to walk on a wooden floor as you do to walk through snow or mud, for example.

- Nerve cells in your leg muscle spindles sense the position of the floor and feedback information to the brain. You do not have to look at the floor to know where it is.

- Once you raise your foot to take a step, your brain sends appropriate signals to the leg and foot muscles to set it down.

This system has sensors (nerve cells, muscle spindles), actuators (muscles) and a controller (brain/spinal cord).

Knowledge of the human healthy and eventually impaired physiology is required to optimally design biomechatronic devices. In particular, biophysical models of muscles, joints, central nervous system and sensors, and human motion control are very helpful for analysis and innovative designs. Also knowledge and skills in mechanical engineering, control engineering, system identification, and signal processing are required to realize devices that improve the quality of life of humans. Example of such devices are deep brain stimulators to suppress the symptoms of Parkinson disease, rehabilitation robotics to enhance neuro-rehabilitation of stroke survivors, wearable exoskeletons for humans that are unable to control their muscles (e.g. Spinal cord injured patients or Duchenne patients), prosthesis, brain computer interfaces, or support of cardiovascular and pulmonary function in the intensive care.

Some Applications

The possibilities of biomechatronics are endless and some of them may be on the border between reality and science fiction but in essence it is a science that can contribute greatly to the improvement of the quality of life of our society.

Here are some of the possibilities that biomechatronics can offer:

- A pancreas pacemakers for diabetics: imagine a device that could sense your sugar blood levels an adjust insulin levels accordingly.

- An electronic muscle stimulators for stroke and accident survivors: people who are in car accidents usually remain disable because of loss of certain muscle functions. Imagine a device that can stimulate damaged muscles and make those muscle contract or relax according to what is needed.

- A digital camera that can be controlled mentally so blind people can get some sort of vision sensation.

- A digital microphone that could be implanted in deaf people so they could hear again or for the first time.

These are some examples were biomechatronics can help.

Modeling of Mechatronic Systems

Goals of Modelling

Models serve many purposes during the design of a mechatronic system. Indeed, the structured analysis and the relevant requirements modelling that has been adopted serves primarily to document and present in a clear fashion the requirements for which a design solution is to be found. Models can also provide a hierarchical framework. Such a framework ensures that the system as a whole is considered but in turn provides the basis of distinguishing the various components of the system which will then allow the division of labour in the design process, permitting concurrent work on separate components of the system. To a certain extent the role of the structured analysis provides such a hierarchical framework through its dataflow diagram levels. Furthermore, models can provide an appropriate insight into the behaviour of a system.

Documentation and Communication

Models present a simplification of the real system and its components and thus provide the basis for summarising a lot of information. Thus, apart from being an effective tool in system analysis, they can be seen as being the best mechanism for communication amongst all people involved in the design and development of the system, from the design team members, to the customers and even to the people involved in the manufacture of the designed system. Indeed, it is possible to identify four distinct groups that stand to benefit from the utilisation of models as communication tools. The engineering team needs communication amongst its members to ensure that effort is the design and development is channelled to the elements and their interoperation that provide the main contribution towards meeting the overall product requirements. In this respect one should include the manufacturing personnel and their contribution towards ensuring a cost effective manufacture of the designed product. Communication with the customer is also critical to ensure final customer satisfaction in the designed and developed system. It is must that through appropriate communication, the customer requirements are correctly understood and implemented in the design. Communication with management is also fundamental in comprehending the status of the project, including the risks undertaken, the costs as well as tradeoffs that would have to be made generally between performance, functionality and cost. Finally, models can serve as a additional tool to store information on the design and developments of products, thus serving as design process documentation which future design teams can benefit from, particularly at earlier design concepts, analysis and decision stages.

When applying models, it should always be remembered that models are a simplification of reality, utilized with a specific scope in mind, for example to understand system information and material flow, or to provide a basis for analyzing system physical dynamics, amongst other aspects. For this purpose, certain models will be suitable for analyzing certain system characteristics, which would be clearly highlighted within the model, while other characteristics will be hidden and ignored within the model. Therefore, if various system characteristics need to be analyzed during the design process, it is likely that various models will be used, each aiding in the analysis of specific distinct characteristics. In this context, as a basis for communication, standardised modelling techniques as well as techniques that are interdisciplinary (and not technology specific) ideally are applied wherever possible.

Hierarchical framework

Mechatronic systems in general include a large number of components that interact in many ways, thus requiring various models as stated earlier on. The various models used would generally aim to providing the means to subdivide complex problems into a set of simpler problems whose solutions can potentially be integrated easily. This role of modelling would thus tend to suggest the use of modelling techniques that allow the subdivision of the model into portions that can be independently tackled by independent engineers. For example, systems of linear differential equations may provide a substantial insight into the fundamental models of a system, but give little insight on how the design problem should be divided among design team members. On the other hand, a model based on dataflow diagrams can provide clear interfaces between elements that allow for system subdivision.

Additionally, when selecting models, consideration has to be additionally given to the mode by which various models can be related together in order to allow cross-referencing between models. When looking at modelling a system, one should therefore also take into consideration the possibility of utilising compatible models or a unifying framework for the models, which will then provide a clearer picture of the general system and its behavior.

Providing Insights into System Behaviour

Regarding insights into system behavior, different models generally provide different types of insight. For example differential equations provide insights regarding the time scales of various behaviors and the relative importance of various factors as well as providing estimates of the validity of simplifying assumptions. Other models can provide insight into the processing power that would be required by the system. In each case the model provides insights into the problem or the characteristics of a proposed solution by bringing into focus certain aspects of the modelled system while hiding others from view.

Physical and Mathematical Modelling of Dynamic Systems

One of the principle scopes for modelling is in the analysis of the system dynamics. A dynamic system can be said to be any collection of interacting elements for which there are cause and effect relationships among the time dependent variables. The analysis of the dynamic behavior of physical systems has become a keystone of modern technology and more than any other field links the different engineering disciplines.

Stages of Dynamic Investigation

The purpose of any dynamic system analysis is to understand and predict the dynamic behavior of a given system and wherever possible to improve upon it. The following defines a list of the steps that one generally goes through during the analysis of dynamic systems:

1. Physical Modelling: where the system to be studied is specified and a simple physical model is adopted, the behavior of which will match sufficiently closely the behavior of the actual system.

2. Equations of Motion: here a mathematical model is derived to represent the physical model, i.e., the differential equations of motion of the physical model are appropriately derived.

3. Dynamic Behavior: the dynamic behavior of the mathematical model can be studied by solving the differential equations of motion.

4. Design Decisions: Following the analysis design decisions can be made, i.e. the physical parameters of the system that give the desired behavior are chosen and implemented within the design as well as the system is augmented, possibly with the introduction of appropriate controllers, so that it will behave as desired.

Physical Modelling: From the Actual System to the Physical Model

A physical model can be considered to be an imaginary physical system which resembles an actual system in its salient features but which is simpler (more "ideal") and is thereby more amenable to analytical studies. When defining a physical model it is necessary to identify which simplifications can be made, ensuring that the model still adequately reflects the actual system. In this respect, engineering judgement plays a fundamental role. Knowing which actual system parameters are likely to have a significant and insignificant effect on the system performance and applying appropriate approximations that will not hinder the correct outcome of the analysis are indeed critical in develop an appropriate physical model. Various approximations are used in physical modelling. A number of these are given in table below.

Table: Typical Approximations in Physical Modelling of Systems

Approximation	Mathematical simplification
Neglect small effects	Reduces number and complexity of differential equations
Assume environment independent of system motions	Reduces number and complexity of differential equations
Replace distributed characteristics with appropriate lumped elements	Leads to ordinary (rather than partial) differential equations
Assume linear relationships	Makes equations linear; allows superposition of solutions
Assume constant parameters	Leads to constant coefficients in differential equations
Neglect uncertainty and noise	Avoids statistical treatment

The table entries are described here under:

Neglect Small Effects: Small effects are neglected on a relative basis. In analyzing the motion of an airplane, we are unlikely to consider the effects of the earth's magnetic field or gravity gradient. To ignore these effects in a space vehicle problem gives grossly incorrect results.

Independent Environment: If we are modelling the influence of the environment on a system component, in reality the system component itself may influence the environment, thus creating a feedback loop. In modelling, such influences are commonly ignored to ensure simplicity of the model. For example, in analyzing the vibration of an instrument panel in a vehicle, we assume that the vehicle motion is independent of the motion of the instrument panel.

Lumped Characteristics: In a lumped-parameter model, system dependent variables are assumed to be constant over finite regions of space rather than over infinitesimal elements, as in a distributed- parameter model. Note that elements in a lumped model do not necessarily correspond to separate physical parts of the actual system. A long electrical transmission line has resistance, inductance, and capacitance distributed continuously along its length. These distributed properties are approximated by lumped elements at discrete points along the line.

Linear Relationships: Nearly all physical elements or systems are inherently nonlinear if there are no restrictions at all placed on the allowable values of the inputs. If the values of the inputs are confined to a sufficiently small range, the original nonlinear model of the system may often be replaced by a linear model whose response closely approximates that of the nonlinear model. When a linear equation has been solved once, the solution is general, holding for all magnitudes of motion. Solutions to linear equations may be superposed on one another.

Constant Parameters: Time-varying systems are ones whose characteristics change with time. Physical problems are simplified by the adoption of a model in which all the physical parameters are constant over time.

Neglect uncertainty and noise: In real systems we are uncertain, in varying degrees, about values of parameters, about measurements, and about expected inputs and disturbances. Disturbances contain random inputs, called noise, which can influence system behavior. It is common to neglect such uncertainties and noise and proceed as if all quantities have definite values that are known.

The most realistic physical model of a dynamic system leads to equations of motion that are; nonlinear, partial differential equations with distributed parameters, with time-varying and space- varying parameters and which include a representation of the uncertain parameters of the system. These equations however are the most difficult to solve and in most practical cases only numerical analysis of the equations is possible. The simplifying assumptions just discussed lead to a physical model of a dynamic system that is less realistic and to equations of motion that are; linear, ordinary differential equations with constant coefficients and without uncertainty factors represented. This type of equation is much easier to solve and the most general and highly developed theory for control design is based on this type of differential equation. However if such models tend to deviate from the real situation, and thus give results of system dynamic behavior that deviate from the real physical system to be designed, then why are such models developed in the first place? The key in this case lies with engineering judgment, i.e. knowing when the model is sufficiently accurate to provide an adequate analysis of the system under investigation. In a number of instances linear ordinary differential equations may suffice in providing an adequate system representation, in other instances it may be necessary to opt for alternative more complex models. It is up to the design engineer to develop a model that can adequately represent reality to an extent that will allow the engineer to derive appropriate design decisions.

Equations of Motion From Physical Model to Mathematical Model

A natural extension of the physical model for analytical purposes is the derived mathematical model, for which a great deal has already been said during the previous discussion on physical models. The focal points in deriving the equations of motion that define the mathematical model for a given physical system model can be stated as follows;

Dynamic Equilibrium Relations: writing of dynamic equilibrium relations to describe the balance of forces, or flow rates, of energy, etc., which must exist for the system and its subsystems. System and subsystem differential equations describe such dynamic equilibria.

Compatibility Relations: writing of system compatibility relations to describe how motions of the system elements are interrelated because of the way they are interconnected. These are inter-element or system relations. Thus transmission ratio and or related motion converter relations relate to such compatibility relations as well as relations related to interconnected electrical and electronic components.

Two further considerations are involved in deriving equations of motion, these being

Physical variables: selection of precise physical variables (velocity, voltage, pressure, flowrate etc.) with which to describe the instantaneous state of a system, and in terms of which to study its behavior.

Physical Laws: natural physical laws which the individual elements of the system obey, including:

- Mechanical relations between force and motion;

- Electrical relations between current and voltage;

- Electromechanical relations between force and magnetic field;

- Thermodynamic relations between temperature, pressure, internal energy, etc..

These relations are called constitutive physical relations as they concern only individual elements or constituents of the system.

Summary of Mathematical Modelling Building Blocks

Table describes the basic building blocks for modelling various systems, including mechanical, electrical, fluid and thermal system elements.

Table: Mathematical modelling building blocks for Physical modelling elements

Physical Element/ Building Block	Modelling Constants/ Variables	Describing Equation	Energy Stored (E)/ Power Dissipated (P)
Mechanical Elements			
Translational – Spring	F = force k = spring constant x = displacement	$F = kx$	$E = \dfrac{1}{2}\dfrac{F^2}{k}$
Translational - Dashpot	F = Force b = damping constant x = displacement v = velocity	$F = b\,\dfrac{dx}{dt}$	$P \quad bv$

Translational – Mass	F = Force m = mass x = displacement v = velocity	$F = m \dfrac{d^2x}{dt^2}$	$E = \dfrac{1}{2}mv^2$
Rotational - Spring	T = torque k = torsional spring const. θ = angular displacement	$T = k\,\theta$	$E = \dfrac{1}{2}\dfrac{T^2}{k}$
Rotational - Damper	T = torque b = torsional damping const. θ = angular displacement ω = angular velocity	$T = b\dfrac{d\theta}{dt}$	$P = b\omega^2$
Rotational –Inertia	T = torque I = moment of inertia θ = angular displacement ω = angular velocity	$T = I\dfrac{d^2\theta}{dt^2}$	$E = \dfrac{1}{2}I\,\omega^2$
Electrical Elements			
Resistance	i = current V = voltage R = resistance	$i = \dfrac{V}{R}$	$P = \dfrac{V^2}{R}$
Inductance	i = current V = voltage L = inductance	$i = \dfrac{1}{L}\int V\,dt$	$E = \dfrac{1}{2}Li^2$
Capacitance	i = current V = voltage C = capacitance	$i = C\dfrac{dV}{dt}$	$E = \dfrac{1}{2}CV^2$
Fluid Power Elements			
Hydraulic Resistance (Resistance to flow)	q = flow rate Δp = pressure difference R = flow resistance*	$q = \dfrac{\Delta p}{R}$	$P = \dfrac{1}{R}\Delta p^2$
Hydraulic Fluid Capacitance (Fluid storage)	q = flow rate Δp = pressure difference C = fluid capacitance $= \dfrac{A}{\rho g}$	$q = C\dfrac{d\Delta p}{dt}$	$E = \dfrac{1}{2}C\Delta p^2$

Hydraulic Fluid Inertance	q = flow rate $\Delta\rho$ = pressure difference I = fluid inertance = $\dfrac{L\rho}{A}$	$q = \dfrac{1}{I}\int \Delta p\, dt$	$E = \dfrac{1}{2}Iq^2$
Pneumatic Resistance	$\dot{m}$ = mass flow rate Δp = pressure difference R = mass flow resistance	$\dot{m} = \dfrac{\Delta p}{dt}$	$P = \dfrac{1}{R}\Delta p^2$
Pneumatic Capacitance	$\dot{m}$ = mass flow rate p = pressure C = pneumatic capacitance = $\dfrac{V}{RT}$ (constant volume)†	$\dot{m} = C\dfrac{dp}{dt}$	$E = \dfrac{1}{2}Cp^2$
Pneumatic Fluid Inertance	$\dot{m}$ = mass flow rate Δp = pressure difference I = fluid inertance = $\dfrac{L}{A}$	$\dot{m} = \dfrac{1}{I}\int \Delta p\, dt$	$E = \dfrac{1}{2}I\dot{m}^2$
Thermal Elements			
Thermal Capacitance	q = heat flow rate T = temperature C = thermal capacitance =Mc	$q = C\dfrac{dT}{dt}$	$E = CT$
Thermal Resistance	q = heat flow rate ΔT = temperature difference R = thermal resistance	$q = \dfrac{\Delta T}{R}$	

General Mathematical Standard Models of Systems

As stipulated earlier on, it is quite common to model systems using standard, ordinary linear differential equations. This is so since a large amount of analysis and a substantial insight into controller development for such system models has been performed on these types of mathematical models. The standard linear ordinary differential equation can be represented as follows;

$$a_n\frac{d^n q_0}{dt^n} + a_{n-1}\frac{d^{n-1} q_0}{dt^{n-1}} + \ldots + a_1\frac{dq_0}{dt} + a_0 q_0 = b_m\frac{d^m q_i}{dt^m} + b_{m-1}\frac{d^{m-1} q_i}{dt^{m-1}} + \ldots + b_1\frac{d^1 q_i}{dt^1} + b_0 q_i$$

In the above equation;

> q_o is the output (response) variable of the physical system
>
> q_i is the input (excitation) variable of the physical system
>
> a_n, and b_m, etc.. are the physical parameters of the system and are considered as constants.

In general it is quite common to try to model systems as either zero, first or second order systems. The equivalent equations for such models are as follows:

Zero-order dynamic system model; $a_0 q_0 = b_0 q_i$

First-order dynamic system model; $a_0 q_0 = b_0 q_i$ $\quad a_i \dfrac{dq_0}{dt} + a_0 q_0 = b_1 \dfrac{dq_i}{dt} + b_0 q_i$

Second-order dynamic system model; $a_2 \dfrac{d^2 q_0}{dt^2} + a_1 \dfrac{dq_0}{dt} + a_0 q_0 = b_2 \dfrac{d^2 q_i}{dt^2} + b_1 \dfrac{dq_i}{dt} + b_0 q_i$

In general, such system models dynamic performance is evaluated through the analysis resulting from specific system inputs q_i. Typical examples in this case would include a step input, ramp input, impulse input as well as a sinusoidal input. Figure illustrates the typical dynamic response curves for a zero order, first-order and second-order system when the input is a step response (i.e. q_i = constant).

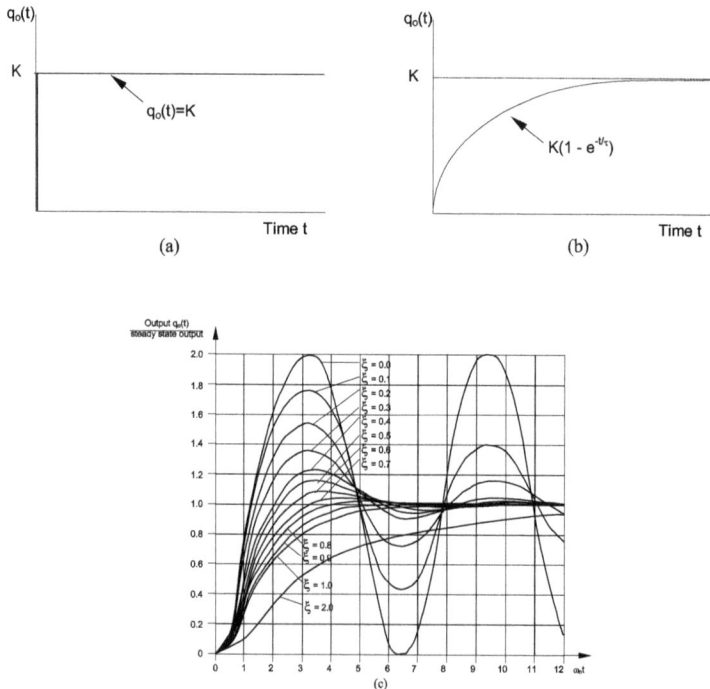

Figure: System step input response for (a) zero-order system, (b) 1st-order system, (c) 2nd-order system.

For the case where the input varies in a sinusoidal manner, system analysis is generally performed over a range of input signal frequencies $q_i = q_{io} \sin(\omega t)$. The resulting output for a sinusoidal input $q_i = q_{io} \sin(\omega t)$, will also be a sinusoidal signal, with the same frequency but with a varying amplitude and phase shift, $q_0 = q_{oo} \sin(\omega t + \phi)$. Thus, the amplitude ratio q_{io} / q_{oo} and phase shift ϕ are generally evaluated with respect to the change in frequency of the input (i.e. the frequency response).

Example of physical and mathematical modeling Electrodynamic vibration shaker

Figure illustrates an electrodynamic vibration shaker. This 'moving coil' type of device converts an electrical signal into a mechanical force and/or motion and is very commonly used in vibration shakers such as vibratory bowl feeders. Here the current-carrying coil is located in a permanent magnetic field which is either generated by a permanent magnet in small devices or by an electrically excited coil in larger applications.

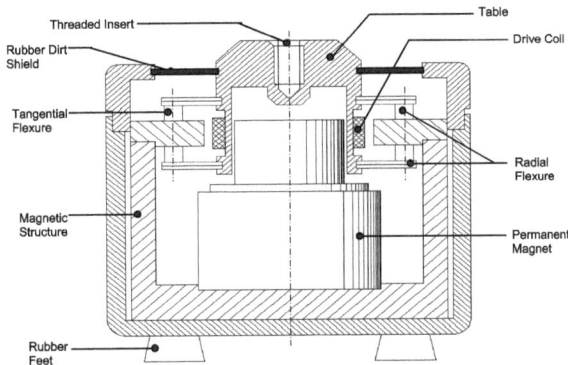

Figure: Schematic of an electrodynamic vibration shaker

Two electromechanical effects are observed in such configurations:

- The motor effect which is the motion of the coil through the magnetic field, causing a voltage proportional to velocity to be induced into the coil.

- The generator effect which is the passage of current through the coil, causing it to experience a magnetic force proportional to the current.

Full-scale force capabilities for different models range from about 20 to 150,000 N with permanent magnet fields being used below about 250N, figure illustrates the physical model applied to the electrodynamic device.

From figure, flexure K_f is an intentional soft spring (yet stiff in the radial direction) that serves to guide the axial motion of coil and table. Flexure damping B_f is usually intentional, fairly strong, and obtained by laminated construction of the flexure spring, using layers of metal, elastomer, plastic, and so on. The coupling of the coil to the shaker table would be rigid so that magnetic force is transmitted undistorted to the mechanical load. Thus K_{tc} (generally large) and B_{tc} (quite small) represent unwanted effects rather than intentional spring and damper elements. R and L are the total circuit resistance and inductance, including contributions from both the shaker coil and the amplifier output circuit.

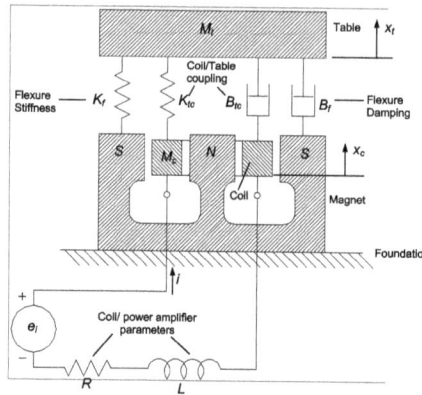

Figure: Physical Model of the electrodynamic vibration shaker

The mathematical model shown hereunder is obtained by application of Newton's law to the table and coil masses and by application of Kirchoff's voltage law to the electrical circuit. The input is the output voltage e_i, of the electronic power amplifier that drives the shaker, and the outputs are the displacements of the table and coil, x_t and x_c, respectively, and the electrical circuit current, i.

$$-K_f x_t - B_f \dot{x}_t - K_{tc}\left(x_t - x_c\right) - B_{tc}\left(\dot{x}_t - \dot{x}_c\right) = M_t \ddot{x}_t$$

$$K_{tc}\left(x_t - x_c\right) + B_{tc}\left(\dot{x}_t - \dot{x}_c\right) + K_{f/i}i = M_c \ddot{x}_c$$

$$-e_i + iR + L\frac{di}{dt} + K_{e/\dot{x}}\dot{x}_c = 0$$

$K_{e/\dot{x}} = K_{f/i}$ represents the electro-mechanical coupling of the coil in V/(m/s) or in N/Amp such that;

Coil back e.m.f. = $K_{e/\dot{x}}\dot{x}_{coil}$

Or Coil force = $K_{f/i}i_{coil}$

From these equations the mathematical model for the system is defined. Through solving such equations, the frequency response of the system can be appropriately evaluated for typical values of the various constant parameters. Thus for various values of the voltage e_i magnitude and frequency, the corresponding vibration oscillations and frequency can be determined by viewing the mode in which the displacement values, x_t in particular, will vary.

Applications of Mechatronics

Mechatronics has a variety of applications as products and systems in the area of 'manufacturing automation'. Some of these applications are as follows:

1. Computer numerical control (CNC) machines

2. Tool monitoring systems

3. Advanced manufacturing systems

   a. Flexible manufacturing system (FMS)

   b. Computer integrated manufacturing (CIM)

Computer Numerical Control Machines

CNC machine is the best and basic example of application of Mechatronics in manufacturing automation. Efficient operation of conventional machine tools such as Lathes, milling machines, drilling machine is dependent on operator skill and training. Also a lot of time is consumed in work part setting, tool setting and controlling the process parameters viz. feed, speed, depth of cut. Thus conventional machining is slow and expensive to meet the challenges of frequently changing product/part shape and size.

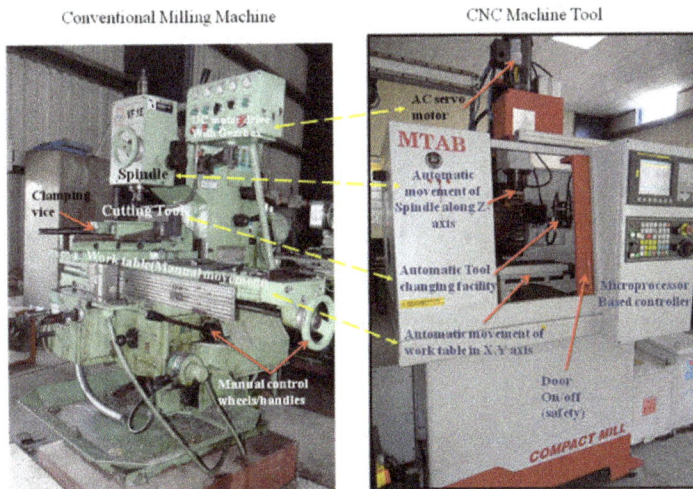

Comparison between a conventional machine tool and a CNC machine tool

Computer numerical control (CNC) machines are now widely used in small to large scale industries. CNC machine tools are integral part of Computer Aided Manufacturing (CAM) or Computer Integrated Manufacturing (CIM) system. CNC means operating a machine tool by a series of coded instructions consisting of numbers, letters of the alphabets, and symbols which the machine control unit (MCU) can understand. These instructions are converted into electrical pulses of current which the machine's motors and controls follow to carry out machining operations on a work piece. Numbers, letters, and symbols are the coded instructions which refer to specific distances, positions, functions or motions which the machine tool can understand.

CNC automatically guides the axial movements of machine tools with the help of computers. The auxiliary operations such as coolant on-off, tool change, door open-close are automated with the help of micro-controllers. Shows the fundamental differences between a conventional and a CNC machine tool. Manual operation of table and spindle movements is automated by using a CNC controllers and servo motors. The spindle speed and work feed can precisely be controlled and maintained at programmed level by the controller. The controller has self-diagnostics facility which regularly alarms the operator in case of any safety norm violation viz. door open during machining, tool wear/breakage etc. Modern machine tools are now equipped with friction-less drives such as re-circulating ball screw drives, Linear motors etc.

Tool Monitoring Systems

Uninterrupted machining is one of the challenges in front manufacturers to meet the production goals and customer satisfaction in terms of product quality. Tool wear is a critical factor which affects the productivity of a machining operation. Complete automation of a machining process realizes when there is a successful prediction of tool (wear) state during the course of machining operation. Mechatronics based cutting tool-wear condition monitoring system is an integral part of automated tool rooms and unmanned factories. These systems predict the tool wear and give alarms to the system operator to prevent any damage to the machine tool and work piece. Therefore it is essential to know how the mechatronics is helping in monitoring the tool wear. Tool wear can be observed in a variety of ways. These can be classified in two groups.

Direct methods	Indirect methods
Electrical resistance	Torque and power
Optical measurements	Temperature
Machining hours	Vibration & acoustic emission
Contact sensing	Cutting forces & strain measurements

Direct methods deal with the application of various sensing and measurement instruments such as micro-scope, machine/camera vision; radioactive techniques to measure the tool wear. The used or worn-out cutting tools will be taken to the metrology or inspection section of the tool room or shop floor where they will be examined by using one of direct methods. However, these methods can easily be applied in practice when the cutting tool is not in contact with the work piece. Therefore they are called as of-fline tool monitoring system. Shows a schematic of tool edge grinding or replacement scheme based on the measurement carried out using offline tool monitoring system. Offline methods are time consuming and difficult to employ during the course of an actual machining operation at the shop floor.

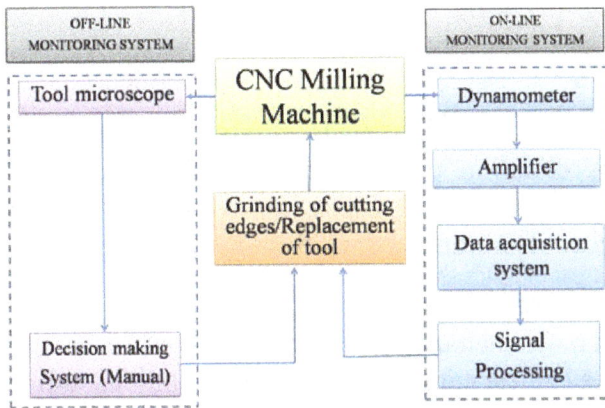

Off-line and on-line tool monitoring system for tool edge grinding

Indirect methods predict the condition of the cutting tool by analyzing the relationship between cutting conditions and response of machining process as a measurable quantity through sensor signals output such as force, acoustic emission, vibration, or current.

Figure shows a typical example of an on-line tool monitoring system. It employs the cutting forces recoded during the real-time cutting operation to predict the tool-wear. The cutting forces can be sensed by using either piezo-electric or strain gauge based force transducer. A micro-processor based control system continuously monitors 'conditioned' signals received from the Data Acquisition System (DAS). It is generally programmed/trained with the past recorded empirical data for a wide range of process conditions for a variety of materials. Artificial Intelligence (AI) tools such as Artificial Neural Network (ANN), Genetic Algorithm (GA) are used to train the microprocessor based system on a regular basis. Based on this training the control system takes the decision to change the tool or gives an alarm to the operator. Various steps followed in On-line approach to measure the tool wear and to take the appropriate action are shown in figure.

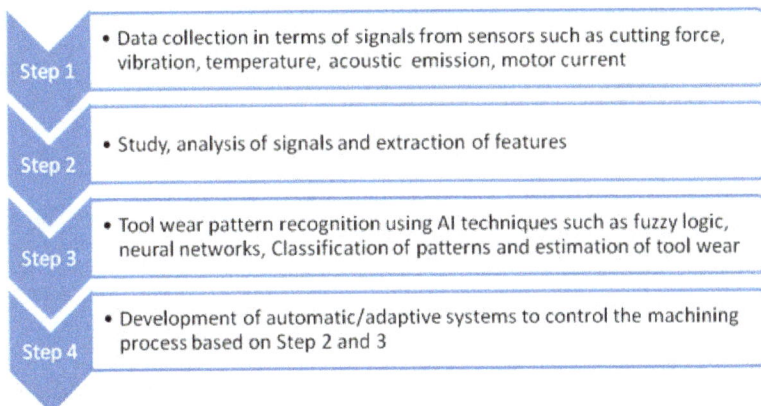

Steps followed in an indirect tool monitoring system

Advanced Manufacturing Systems

Flexible Manufacturing System

Nowadays customers are demanding a wide variety of products. To satisfy this demand, the manufacturers' "production" concept has moved away from "mass" to small "batch" type of production. Batch production offers more flexibility in product manufacturing. To cater this need, Flexible Manufacturing Systems (FMS) have been evolved.

FMS is a manufacturing cell or system consisting of one or more CNC machines, connected by automated material handling system, pick-and-place robots and all operated under the control of a central computer. It also has auxiliary sub-systems like component load/unload station, automatic tool handling system, tool pre-setter, component measuring station, wash station etc. shows a typical arrangement of FMS system and its constituents. Each of these will have further elements depending upon the requirement as given below:

A. Workstations

- CNC machine tools

- Assembly equipment

- Measuring Equipment

- Washing stations

B. Material handing Equipment

- Load unload stations (Palletizing)

- Robotics

- Automated Guided Vehicles (AGVs)

- Automated Storage and retrieval Systems (AS/RS)

C. Tool systems

- Tool setting stations

- Tool transport systems

D. Control system

- Monitoring equipments

- Networks

It can be noticed that the FMS is shown with two machining centers viz. milling center and turning center. Besides it has the load/unload stations, AS/RS for part and raw material storage, and a wire guided AGV for transporting the parts between various elements of the FMS. This system is fully automatic means it has automatic tool changing (ATC) and automatic pallet changing (APC) facilities. The central computer controls the overall operation and coordination amongst the various constituents of the FMS system.

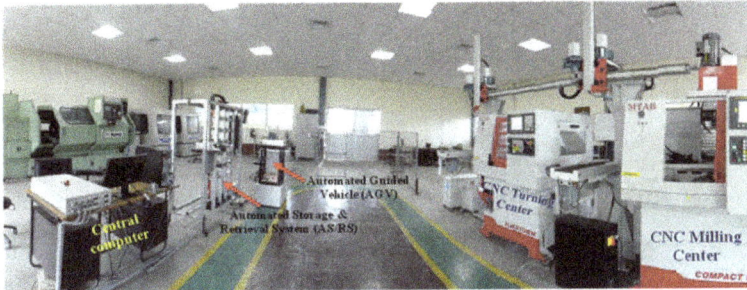

A FMS Setup

The characteristic features of an FMS system are as follows:

1. FMS solves the mid-variety and mid-volume production problems for which neither the high production rate transfer lines nor the highly flexible stand-alone CNC machines are suitable.

2. Several types of a defined mix can be processed simultaneously.

3. Tool change-over time is negligible.

4. Part handling from machine to machine is easier and faster due to employment of computer controlled material handling system.

Benefits of an FMS

- Flexibility to change part variety

- Higher productivity

- Higher machine utilization

- Less rejections

- High product quality

- Reduced work-in-process and inventory

- Better control over production

- Just-in-time manufacturing

- Minimally manned operation

- Easier to expand

CIM is the integration of the total manufacturing enterprise through the use of integrated systems and data communications coupled with new managerial philosophies that improve organizational and personal efficiency'.

CIM basically involves the integration of advanced technologies such as computer aided design (CAD), computer aided manufacturing (CAM), computer numerical control (CNC), robots, automated material handling systems, etc. Today CIM has moved a step ahead by including and integrating the business improvement activities such as customer satisfaction, total quality and continuous improvement. These activities are now managed by computers. Business and marketing teams continuously feed the customer feedback to the design and production teams by using the networking systems. Based on the customer requirements, design and manufacturing teams can immediately improve the existing product design or can develop an entirely new product. Thus, the use of computers and automation technologies made the manufacturing industry capable to provide rapid response to the changing needs of customers.

Application of principles of automation to each of these activities enhances the productivity only at the individual level. These are termed as islands of automation . Integrating all these islands of automation into a single system enhances the overall productivity. Such a system is called as Computer Integrated Manufacturing (CIM).

5

Robotics and Automation

Automation is the technology of performing various operations without human assistance. It can be used for switching on telephone networks, stabilization and steering of aircraft and ships, operating machinery, etc. This chapter investigates the uses of robotics and automation. It includes topics on autonomous car, autonomous logistics, cyborg, laboratory automation, etc.

Automation

Automation is, the application of machines to tasks once performed by human beings or, increasingly, to tasks that would otherwise be impossible. Although the term mechanization is often used to refer to the simple replacement of human labor by machines, automation generally implies the integration of machines into a self-governing system. Automation has revolutionized those areas in which it has been introduced, and there is scarcely an aspect of modern life that has been unaffected by it.

The term automation was coined in the automobile industry about 1946 to describe the increased use of automatic devices and controls in mechanized production lines. The origin of the word is attributed to D.S. Harder, an engineering manager at the Ford Motor Company at the time. The term is used widely in a manufacturing context, but it is also applied outside manufacturing in connection with a variety of systems in which there is a significant substitution of mechanical, electrical, or computerized action for human effort and intelligence.

In general usage, automation can be defined as a technology concerned with performing a process by means of programmed commands combined with automatic feedback control to ensure proper execution of the instructions. The resulting system is capable of operating without human intervention. The development of this technology has become increasingly dependent on the use of computers and computer-related technologies. Consequently, automated systems have become increasingly sophisticated and complex. Advanced systems represent a level of capability and performance that surpass in many ways the abilities of humans to accomplish the same activities.

Automation technology has matured to a point where a number of other technologies have developed from it and have achieved a recognition and status of their own. Robotics is one of these technologies; it is a specialized branch of automation in which the automated machine possesses certain anthropomorphic, or humanlike, characteristics. The most typical humanlike characteristic of a modern industrial robot is its powered mechanical arm. The robot's arm can be programmed to move through a sequence of motions to perform useful tasks, such as loading and unloading parts at a production machine or making a sequence of spot-welds on the sheet-metal parts of an automobile body during assembly. As these examples suggest, industrial robots are typically used to replace human workers in factory operations.

Principles and Theory of Automation

The developments described above have provided the three basic building blocks of automation: (1) a source of power to perform some action, (2) feedback controls, and (3) machine programming. Almost without exception, an automated system will exhibit all these elements.

Power Source

An automated system is designed to accomplish some useful action, and that action requires power. There are many sources of power available, but the most commonly used power in today's automated systems is electricity. Electrical power is the most versatile, because it can be readily generated from other sources (e.g., fossil fuel, hydroelectric, solar, and nuclear) and it can be readily converted into other types of power (e.g., mechanical, hydraulic, and pneumatic) to perform useful work. In addition, electrical energy can be stored in high-performance, long-life batteries.

The actions performed by automated systems are generally of two types: (1) processing and (2) transfer and positioning. In the first case, energy is applied to accomplish some processing operation on some entity. The process may involve the shaping of metal, the molding of plastic, the switching of electrical signals in a communication system, or the processing of data in a computerized information system. All these actions entail the use of energy to transform the entity (e.g., the metal, plastic, electrical signals, or data) from one state or condition into another more valuable state or condition. The second type of action—transfer and positioning—is most readily seen in automated manufacturing systems designed to perform work on a product. In these cases, the product must generally be moved (transferred) from one location to another during the series of processing steps. At each processing location, accurate positioning of the product is generally required. In automated communications and information systems, the terms transfer and positioning refer to the movement of data (or electrical signals) among various processing units and the delivery of information to output terminals (printers, video display units, etc.) for interpretation and use by humans.

Feedback Controls

Feedback controls are widely used in modern automated systems. A feedback control system consists of five basic components: (1) input, (2) process being controlled, (3) output, (4) sensing elements, and (5) controller and actuating devices. These five components are illustrated. The term closed-loop feedback control is often used to describe this kind of system.

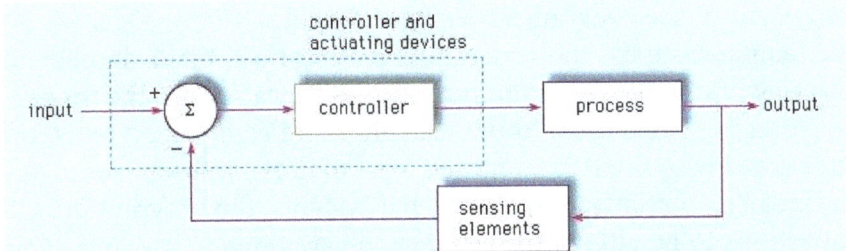

The components of a feedback control system and their relationships.

The input to the system is the reference value, or set point, for the system output. This represents the desired operating value of the output. Using the previous example of the heating system as an illustration, the input is the desired temperature setting for a room. The process being controlled is the heater (e.g., furnace). In other feedback systems, the process might be a manufacturing operation, the rocket engines on a space shuttle, the automobile engine in cruise control, or any of a variety of other processes to which power is applied. The output is the variable of the process that is being measured and compared to the input; in the above example, it is room temperature.

The sensing elements are the measuring devices used in the feedback loop to monitor the value of the output variable. In the heating system example, this function is normally accomplished using a bimetallic strip. This device consists of two metal strips joined along their lengths. The two metals possess different thermal expansion coefficients; thus, when the temperature of the strip is raised, it flexes in direct proportion to the temperature change. As such, the bimetallic strip is capable of measuring temperature. There are many different kinds of sensors used in feedback control systems for automation.

The purpose of the controller and actuating devices in the feedback system is to compare the measured output value with the reference input value and to reduce the difference between them. In general, the controller and actuator of the system are the mechanisms by which changes in the process are accomplished to influence the output variable. These mechanisms are usually designed specifically for the system and consist of devices such as motors, valves, solenoid switches, piston cylinders, gears, power screws, pulley systems, chain drives, and other mechanical and electrical components. The switch connected to the bimetallic strip of the thermostat is the controller and actuating device for the heating system. When the output (room temperature) is below the set point, the switch turns on the heater. When the temperature exceeds the set point, the heat is turned off.

Machine Programming

The programmed instructions determine the set of actions that is to be accomplished automatically by the system. The program specifies what the automated system should do and how its various components must function in order to accomplish the desired result. The content of the program varies considerably from one system to the next. In relatively simple systems, the program consists of a limited number of well-defined actions that are performed continuously and repeatedly in the proper sequence with no deviation from one cycle to the next. In more complex systems, the number of commands could be quite large, and the level of detail in each command could be significantly greater. In relatively sophisticated systems, the program provides for the sequence of actions to be altered in response to variations in raw materials or other operating conditions.

Programming commands are related to feedback control in an automated system in that the program establishes the sequence of values for the inputs (set points) of the various feedback control loops that make up the automated system. A given programming command may specify the set point for the feedback loop, which in turn controls some action that the system is to accomplish. In effect, the purpose of the feedback loop is to verify that the programmed step has been carried out. For example, in a robot controller, the program might specify that the arm is to move to a designated position, and the feedback control system is used to verify that the move has been correctly made. The relationship of program control and feedback control in an automated system is illustrated.

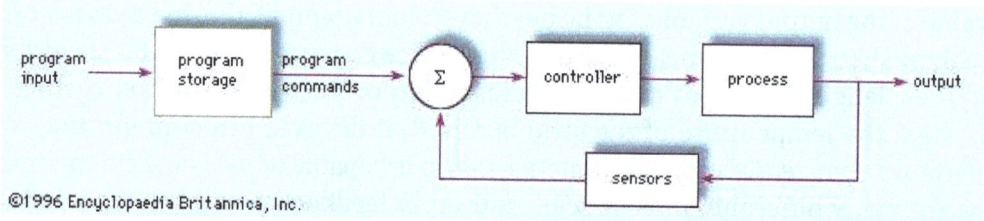

©1996 Encyclopaedia Britannica, Inc.

Relationship of program control and feedback control in an automated system.

Some of the programmed commands may be executed in a simple open-loop fashion—i.e., without the need for a feedback loop to verify that the command has been properly carried out. For example, a command to flip an electrical switch may not require feedback. The need for feedback control in an automated system might arise when there are variations in the raw materials being fed into a production process, and the system must take these variations into consideration by making adjustments in its controlled actions. Without feedback, the system would be unable to exert sufficient control over the quality of the process output.

The programmed commands may be contained on mechanical devices (e.g., mechanical cams and linkages), punched paper tape, magnetic tape, magnetic disks, computer memory, or any of a variety of other media that have been developed over the years for particular applications. It is common today for automated equipment to use computer storage technology as the means for storing the programmed commands and converting them into controlled actions. One of the advantages of computer storage is that the program can be readily changed or improved. Altering a program that is contained on mechanical cams involves considerable work.

Programmable machines are often capable of making decisions during their operation. The decision-making capacity is contained in the control program in the form of logical instructions that govern the operation of such a system under varying circumstances. Under one set of circumstances, the system responds one way; under different circumstances, it responds in another way. There are several reasons for providing an automated system with decision-making capability, including (1) error detection and recovery, (2) safety monitoring, (3) interaction with humans, and (4) process optimization.

Error detection and recovery is concerned with decisions that must be made by the system in response to undesirable operating conditions. In the operation of any automated system, malfunctions and errors sometimes occur during the normal cycle of operations, for which some form of corrective action must be taken to restore the system. The usual response to a system malfunction has been to call for human assistance. There is a growing trend in automation and robotics to enable the system itself to sense these malfunctions and to correct for them in some manner without human intervention. This sensing and correction is referred to as error detection and recovery, and it requires that a decision-making capability be programmed into the system.

Safety monitoring is a special case of error detection and recovery in which the malfunction involves a safety hazard. Decisions are required when the automated system sensors detect that a safety condition has developed that would be hazardous to the equipment or humans in the vicinity of the equipment. The purpose of the safety-monitoring system is to detect the hazard and to take the most appropriate action to remove or reduce it. This may involve stopping the operation and alerting maintenance personnel to the condition, or it may involve a more complex set of actions to eliminate the safety problem.

Automated systems are usually required to interact with humans in some way. An automatic bank teller machine, for example, must receive instructions from customers and act accordingly. In some automated systems, a variety of different instructions from humans is possible, and the decision-making capability of the system must be quite sophisticated in order to deal with the array of possibilities.

A fourth reason for decision making in an automated system is to optimize the process. The need for optimization occurs most commonly in processes in which there is an economic performance criterion whose optimization is desirable. For example, minimizing cost is usually an important objective in manufacturing. The automated system might use adaptive control to receive appropriate sensor signals and other inputs and make decisions to drive the process toward the optimal state.

Example of Automation

Development of Robotics

Robotics is based on two related technologies: numerical control and teleoperators. Numerical control (NC) is a method of controlling machine tool axes by means of numbers that have been coded on punched paper tape or other media. It was developed during the late 1940s and early 1950s. The first numerical control machine tool was demonstrated in 1952 in the United States at the Massachusetts Institute of Technology (MIT). Subsequent research at MIT led to the development of the APT (Automatically Programmed Tools) language for programming machine tools.

A teleoperator is a mechanical manipulator that is controlled by a human from a remote location. Initial work on the design of teleoperators can be traced to the handling of radioactive materials in the early 1940s. In a typical implementation, a human moves a mechanical arm and hand at one location, and these motions are duplicated by the manipulator at another location.

Industrial robotics can be considered a combination of numerical-control and teleoperator technologies. Numerical control provides the concept of a programmable industrial machine, and teleoperator technology contributes the notion of a mechanical arm to perform useful work. The first industrial robot was installed in 1961 to unload parts from a die-casting operation. Its development was due largely to the efforts of the Americans George C. Devol, an inventor, and Joseph F. Engelberger, a businessman. Devol originated the design for a programmable manipulator, the U.S. patent for which was issued in 1961. Engelberger teamed with Devol to promote the use of robots in industry and to establish the first corporation in robotics—Unimation, Inc.

The Robot Manipulator

An industrial robot is a reprogrammable, multifunctional manipulator designed to move materials, parts, tools, or specialized devices through variable programmed motions for the performance of a variety of tasks.

The technology of robotics is concerned with the design of the mechanical manipulator and the computer systems used to control it. It is also concerned with the industrial applications of robots, which are described below.

The mechanical manipulator of an industrial robot is made up of a sequence of link and joint combinations. The links are the rigid members connecting the joints. The joints (also called axes) are the movable components of the robot that cause relative motion between adjacent links. As shown in, there are five principal types of mechanical joints used to construct the manipulator. Two of the joints are linear, in which the relative motion between adjacent links is translational, and three are rotary types, in which the relative motion involves rotation between links.

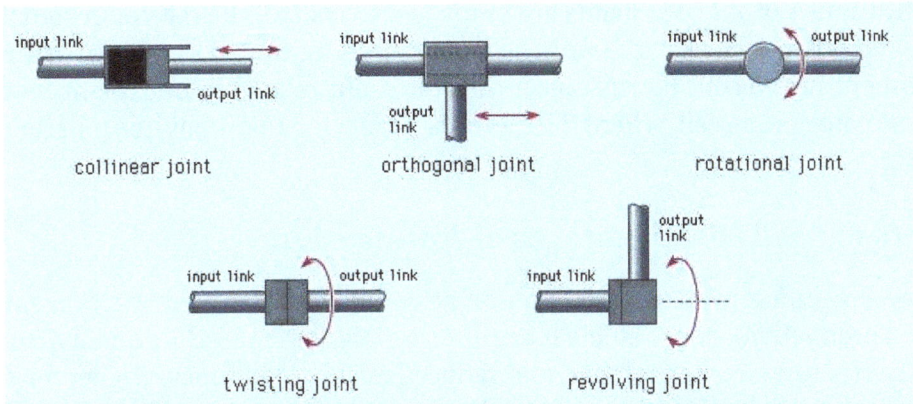

Five types of mechanical joints used in robot manipulators. Collinear and orthogonal
are translational joints; rotational, twisting, and revolving are rotary joints.

The manipulator can be divided into two sections: (1) an arm-and-body, which usually
consists of three joints connected by large links, and (2) a wrist, consisting of two or
three compact joints. Attached to the wrist is a gripper to grasp a work part or a tool
(e.g., a spot-welding gun) to perform a process. The two manipulator sections have
different functions: the arm-and-body is used to move and position parts or tools in
the robot's work space, while the wrist is used to orient the parts or tools at the work
location. The arm-and-body section of most commercial robots is based on one of four
configurations. Each of the anatomies, as they are sometimes called, provides a different work envelope (i.e., the space that can be reached by the robot's arm) and is suited
to different types of applications.

Robot Programming

The computer system that controls the manipulator must be programmed to teach the
robot the particular motion sequence and other actions that must be performed in order
to accomplish its task. There are several ways that industrial robots are programmed.
One method is called lead-through programming. This requires that the manipulator
be driven through the various motions needed to perform a given task, recording the
motions into the robot's computer memory. This can be done either by physically moving the manipulator through the motion sequence or by using a control box to drive the
manipulator through the sequence.

A second method of programming involves the use of a programming language very
much like a computer programming language. However, in addition to many of the
capabilities of a computer programming language (i.e., data processing, computations,
communicating with other computer devices, and decision making), the robot language
also includes statements specifically designed for robot control. These capabilities include (1) motion control and (2) input/output. Motion-control commands are used to
direct the robot to move its manipulator to some defined position in space. For example, the statement "move P1" might be used to direct the robot to a point in space

called P1. Input/output commands are employed to control the receipt of signals from sensors and other devices in the work cell and to initiate control signals to other pieces of equipment in the cell. For instance, the statement "signal 3, on" might be used to turn on a motor in the cell, where the motor is connected to output line 3 in the robot's controller.

Advantages and Disadvantages of Automation

Advantages commonly attributed to automation include higher production rates and increased productivity, more efficient use of materials, better product quality, improved safety, shorter workweeks for labor, and reduced factory lead times. Higher output and increased productivity have been two of the biggest reasons in justifying the use of automation. Despite the claims of high quality from good workmanship by humans, automated systems typically perform the manufacturing process with less variability than human workers, resulting in greater control and consistency of product quality. Also, increased process control makes more efficient use of materials, resulting in less scrap.

Worker safety is an important reason for automating an industrial operation. Automated systems often remove workers from the workplace, thus safeguarding them against the hazards of the factory environment. In the United States the Occupational Safety and Health Act of 1970 (OSHA) was enacted with the national objective of making work safer and protecting the physical well-being of the worker. OSHA has had the effect of promoting the use of automation and robotics in the factory.

Another benefit of automation is the reduction in the number of hours worked on average per week by factory workers. About 1900 the average workweek was approximately 70 hours. This has gradually been reduced to a standard workweek in the United States of about 40 hours. Mechanization and automation have played a significant role in this reduction. Finally, the time required to process a typical production order through the factory is generally reduced with automation.

A main disadvantage often associated with automation, worker displacement, has been discussed above. Despite the social benefits that might result from retraining displaced workers for other jobs, in almost all cases the worker whose job has been taken over by a machine undergoes a period of emotional stress. In addition to displacement from work, the worker may be displaced geographically. In order to find other work, an individual may have to relocate, which is another source of stress.

Other disadvantages of automated equipment include the high capital expenditure required to invest in automation (an automated system can cost millions of dollars to design, fabricate, and install), a higher level of maintenance needed than with a manually operated machine, and a generally lower degree of flexibility in terms of the possible products as compared with a manual system (even flexible automation is less flexible than humans, the most versatile machines of all).

Also there are potential risks that automation technology will ultimately subjugate rather than serve humankind. The risks include the possibility that workers will become slaves to automated machines, that the privacy of humans will be invaded by vast computer data networks, that human error in the management of technology will somehow endanger civilization, and that society will become dependent on automation for its economic well-being.

These dangers aside, automation technology, if used wisely and effectively, can yield substantial opportunities for the future. There is an opportunity to relieve humans from repetitive, hazardous, and unpleasant labour in all forms. And there is an opportunity for future automation technologies to provide a growing social and economic environment in which humans can enjoy a higher standard of living and a better way of life.

Autonomous Car

A self-driving car (sometimes called an autonomous car or driverless car) is a vehicle that uses a combination of sensors, cameras, radar and artificial intelligence (AI) to travel between destinations without a human operator. To qualify as fully autonomous, a vehicle must be able to navigate without human intervention to a predetermined destination over roads that have not been adapted for its use.

Companies developing and/or testing autonomous cars include Audi, BMW, Ford, Google, General Motors, Tesla, Volkswagen and Volvo. Google's test involved a fleet of self-driving cars -- including Toyota Prii and an Audi TT -- navigating over 140,000 miles of California streets and highways.

Levels of Autonomy in Self-driving Cars

The U.S. National Highway Traffic Safety Administration (NHTSA) lays out six levels of automation, beginning with zero, where humans do the driving, through driver assistance technologies up to fully autonomous cars. Here are the five levels that follow zero automation:

- Level 1: Advanced driver assistance system (ADAS) aid the human driver with either steering, braking or accelerating, though not simultaneously. ADAS includes rearview cameras and features like a vibrating seat warning to alert drivers when they drift out of the traveling lane.

- Level 2: An ADAS that can steer and either brake or accelerate simultaneously while the driver remains fully aware behind the wheel and continues to act as the driver.

- Level 3: An automated driving system (ADS) can perform all driving tasks under certain circumstances, such as parking the car. In these circumstances, the human driver must be ready to re-take control and is still required to be the main driver of the vehicle.

- Level 4: An ADS is able to perform all driving tasks and monitor the driving environment in certain circumstances. In those circumstances, the ADS is reliable enough that the human driver needn't pay attention.

- Level 5: The vehicle's ADS acts as a virtual chauffeur and does all the driving in all circumstances. The human occupants are passengers and are never expected to drive the vehicle.

As of 2018, car makers have reached Level 3. Self-driving vehicles are at least a few years off, because manufacturers must clear a variety of technological milestones and a number of important issues must be addressed before autonomous vehicles can be purchased and used on public roads in the United States, according to the NHTSA.

How Self-driving Cars Work

AI technologies power self-driving car systems. Developers of self-driving cars use vast amounts of data from image recognition systems, along with machine learning and neural networks, to build systems that can drive autonomously.

The neural networks identify patterns in the data, which is fed to the machine learning algorithms. That data includes images from cameras on self-driving cars from which the neural network learns to identify as traffic lights, trees, curbs, pedestrians, street signs and other parts of any given driving environment.

For example, Google's self-driving car project, called Waymo, uses a mix of sensors, Lidar (light detection and ranging -- a technology similar to radar), and cameras and combines all of the data those systems generate to identify everything around the vehicle and predict what those objects might do next. This happens in fractions of a second. The system learns more as it drives, so maturity is important with these systems.

How Google Waymo vehicles work:

- The driver (or passenger) sets a destination. The car's software calculates a route.

- A rotating, roof-mounted Lidar sensor monitors a 60-meter range around the car and creates a dynamic 3D map of the car's current environment.

- A sensor on the left rear wheel monitors sideways movement to detect the car's position relative to the 3D map.

- Radar systems in the front and rear bumpers calculate distances to obstacles.

- AI software in the car is connected to all the sensors and collects input from Google Street View and video cameras inside the car.

- The AI simulates human perceptual and decision-making processes and controls actions in driver-control systems such as steering and brakes.

- The car's software consults Google Maps for advance notice of things like landmarks and traffic signs and lights.

- An override function is available to allow a human to take control of the vehicle.

The Pros and Cons of Autonomous Cars

The top benefit touted by autonomous vehicle proponents is safety. A U.S. Department of Transportation and NHTSA statistical projection of traffic fatalities for 2017 estimated that 37,150 people died in motor vehicle traffic crashes that year. The NHTSA estimates that 94% of serious crashes are due to human error or poor choices, such as drunk or distracted driving. Autonomous cars remove those risk factors from the equation -- though self-driving cars are still vulnerable to other factors, such as mechanical issues, that cause crashes.

If autonomous cars can significantly reduce the number of crashes, the economic benefits could be enormous. Injuries impact economic activity, including $57.6 billion in lost workplace productivity, and $594 billion due to loss of life and decreased quality of life due to injuries, according to the NHTSA.

In theory, if the roads were mostly occupied by autonomous cars, traffic would flow smoothly and there would be less traffic congestion. In cars that are fully automated, the occupants could do productive activities while commuting to work. People who aren't able to drive due to physical limitations could find new independence through autonomous vehicles and would have the opportunity to work in fields that require driving.

Autonomous trucks have been tested in the U.S. and Europe to allow drivers to use autopilot over long distances, freeing the driver to rest or complete tasks and improving driver safety and fuel efficiency. This initiative, called truck platooning, is powered by adaptive cruise control (ACC), collision avoidance systems and vehicle-to-vehicle communications for cooperative adaptive cruise control (CACC).

The downsides of autonomous vehicle technology could be that riding in a vehicle without a driver behind the steering wheel may be unnerving, at least at first. But as self-driving capabilities become commonplace, human drivers may become overly reliant on the autopilot technology and leave their safety in the hands of automation, even when they should act as backup drivers in case of software failures or mechanical issues.

In one example from March 2018, Tesla's Model X SUV was on autopilot when it crashed into a highway lane divider. The driver's hands were not on the wheel, despite visual warnings and an audible warning to put his hands back on the steering wheel, according to the company.

Autonomous Car Safety and Challenges

Autonomous cars must learn to identify countless objects in the vehicle's path, from branches and litter to animals and people. Other challenges on the road are tunnels that interfere with Global Positioning Systems (GPS), construction projects that cause lane changes, or complex decisions, like where to stop to allow emergency vehicles to pass.

The systems need to make instantaneous decisions on when to slow down, swerve or continue acceleration normally. This is a continuing challenge for developers and there are reports of self-driving cars hesitating and swerving unnecessarily when objects are detected in or near the roadways.

This problem was clear in a fatal accident in March 2018 which involved an autonomous car operated by Uber. The company reported that the vehicle's software identified a pedestrian but deemed it a false positive and failed to swerve to avoid hitting her. This crash caused Toyota to temporarily cease its testing of self-driving cars on public roads, but its testing will continue elsewhere. The Toyota Research Institute is constructing a test facility on a 60-acre site in Michigan to further develop automated vehicle technology.

With crashes also comes the question of liability, and lawmakers have yet to define who is liable when an autonomous car is involved in an accident. There are also serious concerns that the software used to operate autonomous vehicles can be hacked, and automotive companies are working to address cyber security risks.

Car makers are subject to Federal Motor Vehicle Safety Standards, and the NHTSA reports that more work must be done for vehicles to meet those standards.

The Road to Driverless Cars

The path toward self-driving cars began with incremental automation features for safety and convenience before the year 2000, with cruise control and antilock brakes. After the turn of the millennium, advanced safety features including electronic stability control, blind spot detection and collision and lane shift warnings became available in vehicles. Between 2010 and 2016, advanced driver assistance capabilities such as rearview video cameras, automatic emergency brakes and lane centering assistance emerged, according to the NHTSA.

Since 2016, automation has moved toward partial autonomy, with features that help drivers stay in their lane, along with adaptive cruise control technology, and the ability to self-park.

Fully automated vehicles are not publicly available yet and may not be for many years. In the U.S., the NHTSA provides federal guidance for introducing automated driving systems onto public roads and as autonomous car technologies advance, so will the department's guidance.

Self-driving cars are not yet legal on most roads. In June 2011, Nevada became the first jurisdiction in the world to allow driverless cars to be tested on public roadways; California, Florida, Ohio and Washington, D.C., have followed in the years since.

Autonomous Logistics

The modern logistics industry is becoming a hub for exciting robotic applications. It was slow to adopt automation at first, but the last few years have seen a rise in robotic solutions to logistics problems. According to Machine Handling and Logistics, "Some of the largest [logistics] operators in the US have plans to automate almost every physical move in their facilities within the next two to three years".

These applications range from (somewhat mundane) palletizing tasks all the way up to self-driving trucks and delivery drones.

Here are ten of the most interesting emerging applications in autonomous logistics today.

Warehousing Robots

One of the most mature robotic technologies in logistics is autonomous warehousing. Since Amazon acquired Kiva Systems in 2012, robotic warehouses have grown in popularity. They reduce the time that workers spend traveling around the warehouse by having robots bring shelves to the workers instead.

Although Amazon was one of the first big businesses on the scene, their robotic warehouses certainly aren't the only ones. In May 2018, logistics giant Rakuten Super Logistics partnered with InVia Robotics to automate their US warehouses, and more partnerships are surely on the way.

Autonomous Picking

Most warehousing robots these days work by bringing shelves to human workers, who then pick items by hand. However, more and more robots are able to pick items from static shelves without human assistance. This involves grasping diversely shaped items—a complex task. It would be great for logistics companies if all robots were capable of this, which is why Amazon hosts its annual Picking Challenge.

Robotic Packing

Packing is one of the most common logistics tasks—after all, everything shipped must be packed. Robots can be used for many of the several steps in packaging. At first, products are packed into individual packages. This is often done by the manufacturer. The next two steps are additional packing into larger boxes and crates; the latter step, called tertiary packaging, is often carried out by the logistics company.

Packing is one of the most common tasks in logistics.

Robot Palletizing

One example of tertiary packaging is palletizing. It's well suited to robots because it generally involves repetitive actions and items that are too heavy for humans to lift.

There are even specialized palletizing robots: these are usually robotic arms that have five degrees of freedom (DoF) and a sixth DoF that can rotate more than 360 degrees, so that the robot can properly orient items onto pallets. There are also many programming solutions to simplify robot palletizing, such as the UR palletizing wizard.

Robotic Ports

One of the newest robotic-logistic applications is robotic ports. These aren't USB or Ethernet ports—they're where cargo is loaded onto boats.

Robotic ports were first heard about back in May when a Chinese port was made fully robotic. The port was designed by TuSimple, a self-driving-truck startup. It uses autonomous cranes and self-driving container vehicles to perform all the functions of a cargo port, but with no humans on site.

Self-driving Vehicles

With the rapid advances made in the past few years, autonomous trucks are the most obvious example of self-driving technology. Other applications include autonomous forklifts, delivery vehicles, and mobile parcel stations (like mailboxes on wheels).

As with the TuSimple port, China seems to be a driving force (if you'll excuse the pun) for autonomous logistics vehicles. "Dozens of autonomous truck startups are reported to have launched in China in the past two years".

Last Mile Drone Delivery

Last mile delivery is (as the name suggests) the last step in the logistics journey from manufacturer to end customer. Although drone delivery has received a lot of media attention, drones aren't the only robots being tested at this stage. Other recent developments include autonomous ground, super-size, and four-legged delivery robots.

Container Loading and Unloading

Container loading is like palletizing, but instead of piling products onto a pallet, it involves pulling packages out of a shipping container. As a result, the handling requirements are slightly different, since the robot must be able to hold products from the side. One example of such a system is the IRIS project, which seeks to use vision and self-learning to improve packing.

Sorting

Sorting is essential at many stages of logistics. But it has traditionally been challenging for robots, which require advanced sensing to differentiate between various products. However, the necessary capabilities are easier to acquire than ever before thanks to vision sensors like the Robotiq Wrist Camera.

Collaborative Robot Packing

As you know if you've been following the Robotiq blog, cobots are on the rise in many industries, and logistics is no different. Collaborative packing lets humans and cobots work together: the robot can take over the more mundane, repetitive actions, while humans handle the more delicate tasks.

Autonomous Spaceport Drone Ship

SpaceX recently managed to land its Falcon 9 reusable rocket on its autonomous drone ship out at sea after four unsuccessful attempts, with plans for at least two more sea landings later this year. It takes a special kind of ship to catch a rocket falling from the sky, and here's how they do it.

Elon Musk just had the best week of business in his life after finally successfully landing the SpaceX Falcon 9 rocket engine on a tiny speck of a refitted cargo barge floating in the ocean, in addition to the 325,000 reservations for the Tesla Model 3 electric sedan.

But the riskiest venture of the week was the final moment the Falcon 9 rocket engine returned to Earth after launching an addition to the International Space Station into orbit and, for the first time, managed to land upright on SpaceX's autonomous landing ship, the wonderfully-named "Of Course I Still Love You".

It turns out that the landing platforms that SpaceX has developed are just as complicated as literal rocket science.

There have been three separate landing ships developed by SpaceX. The concept of an "Autonomous Spaceport Drone Ship" harks back to Elon Musk's plans for "creating a paradigm shift in the traditional approach for reusing rocket hardware" in 2009, but the contract for a refitted barge was officially announced in 2014.

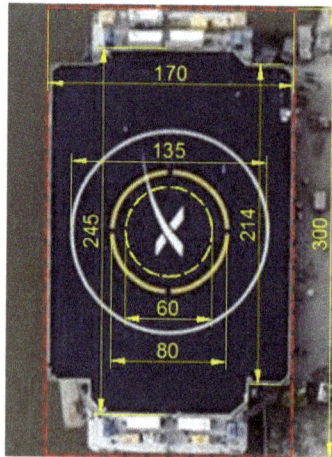

Photo originally tweeted (and since deleted) by Elon Musk of the landing platform

The large, flat surface of the ocean barge and the ability to transport heavy payloads are necessary for the high-velocity landings of the Faclon 9 rocket. SpaceX refits the barge platforms with an extended platform measuring 170 feet by 300 feet to facilitate the 18 foot span of the landing legs and the rocket's support supplies. The first ship featured extended wings and blast walls, but the current models no longer have these features.

According to NASA, these Marmac autonomous drone ships are outfitted with four diesel-powered azimuth thruster engines which can pivot horizontally, removing the need for a ship rudder and offering better maneuverability with a complete 360 degree range of motion for the ship controls. It was originally reported that these thrusters help hold the drone ship position within three meters of accuracy. The engines are supplied by marine equipment manufacturer Thrustmaster.

The materials on the surface of the ship aren't entirely known, but is likely a thick sheet of steel which prominently features the SpaceX logo.

The drone ships are capable of maneuvering autonomously, using GPS information for precision positioning, but they can also be remotely controlled by an accompanying support ship with a crew of technicians standing by.

A wide array of sensors are used in tandem with the GPS information to manage the "attitude and placement" of the drone ship and communicate with the incoming rocket and the onlooking teams to control and coordinate the complicated precise landings.

Following each landing attempt the crews of the support ship board the drone ship. If the landing is successful, the standing rocket is welded to the deck of the ship and reinforced until it arrives back in port. If the landing was a failure, whatever is salvageable is secured and returned to port.

The Atlantic drone ship returns to the Port of Jacksonville or Port Canaveral in Florida, and the Pacific drone ship docks in the Port of Los Angeles in California.

The first "Just Read The Instructions," which was retired after six months of service. Photo: SpaceX

The "Of Course I Still Love You," which received the first successful Falcon 9 landing April 8th, 2016. Photo: SpaceX

The first implemented use of SpaceX's Autonomous Spaceport Drone Ship was in the Atlantic Ocean following a cargo launch to the International Space Station for NASA on January 10th, 2015, which resulted in a failed landing attempt.

Elon Musk claims the liklihood of a successful landing was 50 percent at the most, but recently stated that the odds are closer to 30 percent.

The first drone ship was based on a Marmac series 300 ocean barge, and its name was revealed to be "Just Read The Instructions" after its first test landing.

The first drone ship "Just Read The Instructions" was replaced for Atlantic Ocean duties after six months of operations and two failed test landings at sea.

It's replacement platform was leased by SpaceX and developed on a much newer Marmac barge, model 304, and as I mentioned earlier, named "Of Course I Still Love You."

Both ship titles are creatively named after autonomously-controlled spaceships from the novel The Player of Games.

A third floating landing platform based on the leased Marmac barge model number 303 was finished in the spring of 2015 and transited the Panama Canal for Pacific Ocean rocket recovery duties. It was later named "Just Read The Instructions" like the first, now retired floating landing platform.

SpaceX has attempted to land the Falcon 9 rocket engine on an autonomous drone ship five times. Two landing attempts in January and April of 2015 resulted in failures after contact with the drone ship, but many stages of the attempts were successful.

The third intended landing attempt was thwarted in June 2015 after the rocket disintegrated before entering the first stage of the landing process.

The first Pacific landing attempting, and fourth total, was conducted in January with a successful "soft landing" of the rocket on the drone ship. However, a catch pin in one of the rocket's landing legs failed and the rocket toppled over.

The fifth landing attempt on Friday was the first full success in SpaceX's attempts to land its Falcon 9 rocket engine on a drone ship at sea. This success, and the hope for many more like it in the future, mark the underlying purpose behind SpaceX's reusable rocket program.

The ability to completely recover an entire rocket body and return it to port to prep for reuse will drastically cut the cost of spaceflight and increase the pace at which payloads and, eventually, humans will be able to access space.

According to Elon Musk at SpaceX's press conference following the successful landing, eventual use of the drone ships should drop to a quarter of SpaceX return landings with a shift towards land locations for returning rockets. Elon Musk also once tweeted an eventual possibility where the rocket engines land on the drone ship, are prepped and refueled in a rapid turnaround time, and then fly themselves "home".

Musk also brought up plans for the first manned SpaceX flight by the end of next year with the second generation of the Dragon capsule, with an unmanned test planned first.

"We'll be successful when it's boring," Musk quipped at a question asking him about the next goal following this major success. But he has a point—the realization that this sort of marveling engineering achievement brings us one step closer to casual, inexpensive spaceflight is truly sobering.

The ability for a rocket delivering payloads beyond our atmosphere to communicate and successfully connect with an autonomous floating platform in rocky seas not only breaks the traditional mold of spaceflight recovery, but almost literally jumps from the pages of science fiction.

Cyborg

A cybernetic organism or "cyborg" in IT is defined as an organism with both biological and technological components. In some definitions, a cyborg is described as a hypothetical or fictional creation. However, in a technical sense, humans can be seen as cyborgs in various types of situations, including the use of artificial implants.

Part of the diverse use of the word "cyborg" revolves around how humans see their interactions with technology. A person could be considered a cyborg when they are outfitted with implants such as artificial heart valves, cochlear implants or insulin pumps. A person could even be called a cyborg when they are using specific wearable technologies like Google Glass, or even using laptops or mobile devices to do work.

However, a different definition of a cyborg involves fictional pictures of human individuals with enhanced virtual-reality vision, robotic implants on limbs and torso, and other more significant body IT components. The popular definition of cyborg changes as a range of science-fiction-type ideas become realities.

Laboratory Automation

Laboratory testing has grown from a manual, "hands-on" process providing a simple test menu to an instrument-centric, high-volume clinical engine inside the modern healthcare enterprise. While the instruments have grown both in size and scale, the proverbial "glue" that holds the laboratory together is the lines of automation. These "conveyor belts" for laboratory samples bring fast and accurate routing of specimens to specific points in the laboratory work flow. Because of this growth, automation has moved past the "nice to have" to the "must have" in the modern laboratory, and automation solutions are beginning to extend their footprints into other areas of the laboratory such as microbiology and molecular diagnostics.

With the three components of automation growing within the clinical laboratory in the late 1980s (instrument-level automation, the LIS/LIMS, and early pre- and post-analytic automation), the strategies for laboratory automation have taken a variety of routes. First, the degree of automation at the level of the instrument has grown to handle higher throughputs as the demands of the laboratory have grown. This growth is easily seen in the orders of magnitude increases in instrument throughput over the past few decades. The LIS/LIMS systems have grown from simple appliances that handle the mechanics of laboratory billing to complete, end-to-end platforms that govern the entire business process and work flow of the laboratory. These IT systems are now mission-critical to the operation of a laboratory and must work in concert with instruments and automation for a laboratory to be successful.

The pre- and post-analytic automation solutions are present in one of two varieties: open and closed. Of the two, closed automation is the more common. Closed automation solutions are those that are provided by instrument manufacturers and typically connect only to instruments from that vendor. And unless a single vendor has instruments in multiple parts of the lab, closed automation solutions exist as islands of automation around their specific parts of the lab. Open automation solutions exist independently from the instruments in the laboratory. These solutions are built by independent companies and interface to the instruments and the LIS/LIMS to automate the pre- and post-analytic work flows.

The main difference between the open and closed approaches is the types of instruments connected to the automation line. Unlike the closed lines that are provided with instruments, the open automation solutions are designed and acquired by labs independently of the instruments in the laboratory. While the open automation lines connect to instruments in a similar fashion (direct track sampling or robotic arm), open automation lines can cross service areas of the lab because of the ability to interface to any instrument, regardless of the vendor. This independence from a specific vendor allows open automation systems to transport specimens across various parts of the lab and provide a single automation solution for the entire laboratory.

The current "whys" for automation in the laboratory are error reduction and staff augmentation. Both are driven by increased testing demands and laboratory staffing needs. Automation has indeed moved from "nice to have" for large reference laboratories to "must have" for any clinical laboratory. The current climate for reimbursements for laboratory testing has only increased the importance of automation, with further reductions that push laboratories to optimize laboratory throughput and staffing to keep up with the increasing demands.

Benefits of Lab Automation

Provides Valuable Walk-away Time

There are many important tasks to get done in the lab, so why not eliminate the tedious and time consuming task of hand labeling tubes? Imagine how much time you will save by giving this job to a fully automated labeling machine that can fill, read 2D barcode for verification, and place your tubes back into the rack. Labeling tubes is a rather simple task, so is it really necessary to have a scientist perform such tedious work when he or she could be using their valuable skills for a more specialized task.

Reduces Repetitive Injuries

Manual labeling of tubes is not only tedious and time-consuming but it can also lead to physical strain of the fingers and hands. People who have been hand labeling tubes for years might start to experience numbness and tingling in the hand, which ultimately decreases productivity in the lab. If your daily routine consists of labeling, filling, un-capping and capping of tubes, this can cause stress injuries in the fingers and in many cases lead to carpal tunnel syndrome. Many fully-automated labeling machines can complete all of these tasks for you and decrease the risk of repetitive injuries from manual labeling.

Reduces Costs in the Lab

Using laboratory automation in your lab will decrease the number of lab techni-cians needed to label tubes, therefore resulting in cost savings. Not only will you reduce costs in the lab, but you also might find an increase in revenue due to lab automation.

Eliminates Human Error

There are many steps involved in preparing and keeping track of samples in the lab. Lab automation not only decreases human error associated with sample preparation but also it increases efficiency in the lab by allowing the user to track samples. If you are working with a LIMS system or excel work list, these can easily be imported into the software to ensure tracking of samples, therefore reducing the amount of lost or misplaced samples.

Flexible Machinery Saves Valuable Space

Saving space in the lab is a legitimate concern. Although some lab automation systems have a large footprint and take up lots of space, there are also many benchtop systems that are designed to fit easily in a lab work space.

Laboratory Robotics

Laboratory robotics is the practice of using robots to perform or assist in laboratory tasks. While laboratory robots have found applications in diverse industries and sci-ences, pharmaceutical companies have used them more than any other group.

Advantages of laboratory robotics include high speed, high efficiency, minimal wast-age, task reproducibility, task endurance, precision, high productivity, enhanced safety for laboratory personnel, the ability to withstand adverse environmental conditions, and reduced tedium and boredom among laboratory workers who would otherwise perform the tasks.

Laboratory robotics has some shortcomings and limitations such as high initial cost, limited flexibility, and lack of intuition for solving difficult or esoteric problems. In addition, layoffs of human workers (as robots replace them) can damage employee mo-rale.

References

- Automation, technology: britannica.com, Retrieved 17 April 2018

- Advantages-and-disadvantages-of-automation, technology: britannica.com, Retrieved 16 June 2018

- Driverless-car: searchenterpriseai.techtarget.com, Retrieved 28 May 2018

- Spacexs-landing-drone-ship-is-just-as-complicated-as-th-1769987148:   foxtrotalpha.jalopnik.com, Retrieved 10 April 2018

- Cyborg-15651: techopedia.com, Retrieved 30 June 2018

- 5-benefits-lab-automation: scinomix.com, Retrieved 09 July 2018

- Laboratory-robotics: searchenterpriseai.techtarget.com, Retrieved 27 May 2018

Other Aspects of Robotics

The fields of mechatronics and robotics have witnessed significant advancement in the past few decades owing to progress in science and technology. The topics elucidated in this chapter address the varied robotic systems, such as care-O-bot, bowler communication system, etc. as well as discuss about real-time path planning, robot interaction language, perceptual control matter, etc.

Bin Picking

Bin picking is the name of the technique that is being used by a robot to grab objects that are randomly placed inside a box or on a pallet. Bin picking allows a robot to grab objects rapidly without the need for extensive programming-time. It is a big improvement in efficiency and probably the biggest challenge we currently face in the field of robot technology.

For humans, bin picking is daily practice. We are able to move objects, sort them and stack them without even thinking about it. For example, we are able to pick a peach out of a box with apples and cleaning a desk is a simple task for us. We automatically apply the right force, using just our brains.

However, all the above situations are very difficult for a robot to process. Normally it would take an extreme amount of programming time to set up a robot to do such tasks. This is the reason Teqram is developing a solution for this problem. This solution is called "bin picking".

Bin picking is about grabbing unsorted objects by a robot. In cooperation with clever vision and sensor systems, the robot is able to scan the surface for products to pick and move to the next location.

Robots are equipped with sensors. Every sensor passes their own information, but often the robot is still restricted in what it feels and sees. Originally, robot setups would be programmed for repeating tasks. Programming time would thus be spread out among many products. These robots are not 'smart' because engineers tell them what to do by programming.

This all changes with the bin picking technique. Tasks are hardly ever the same. Products are scattered randomly inside a box or on a pallet. A robot has to be equipped with the right sensors to differentiate products. Next, the robot has to move to the object without damaging surrounding objects. This process would need a revolution in robot control.

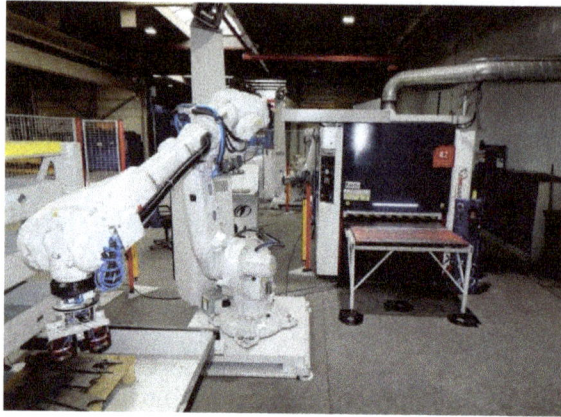

Structured vs Random Bin Picking

random bin pickingRobotic bin picking is a term used to describe a variety of processes where a robot is used to pick and place components from a bin. Depending on what happens upstream in the component manufacturing, the final bins can be setup in a variety of ways.

Structured Bin Picking

If in the upstream process, parts are placed into a bin in an organized, predictable pattern it is called structured bin picking.

Fairly simple and straightforward robotic application that has been mainstream in facilities for many years:

- Requires more work at upstream process to organize parts.

- Typically is least expensive.

- Dunnage may or may not be required for layers and/or parts.

- Can simplify or eliminate the need for expensive vision process to locate and inspect parts.

- Vision inspection of some or 100% of parts can be accomplished.

- Likely will have fastest part to part cycle times.

Semi – Structured Bin Picking

The next level is a Semi-Structured bin where there is some predictability in part placement or parts are separated so that they can more easily be imaged and picked. This presents a slightly more difficult application for the robot, and it also encompasses a diverse amount of situations.

- Requires work at upstream process either to layer or place parts.

- Mid-range expense depending on actual part arrangement.

- Dunnage may or may not be required for layers and/or parts.

- Simplifies the vision process to locate and inspect parts.

- Vision inspection is an option.

- Part to part cycle times again range based on the level or structure.

Random Bin Picking

Lastly, there is the emerging application of completely unorganized parts that are thrown or dumped (bulk loaded) into a bin upstream in the process. Each bin that arrives at the robot cell contains components that are arranged differently and part orientation is completely unpredictable. This type of bin picking is commonly known as the "Holy Grail" in robotic applications. Technology has changed this perception. Random-robotic-bin-pickingThe increase in camera resolution, processing power, and EOATs sensitivity and flexibility as allowed some robot integrators to meet this complicated task of picking randomly ordered parts from a bin.

- Save money upstream by being able to randomly place parts in a bin.

- New vision and end of arm tooling technologies are used to locate and pick parts.

- Part to part cycle times are longer but can be optimized to meet rate needs.

Bowler Communications System

BowlerStudio is a robot development application that combines scripting and device management with powerful control and processing features.

The goal is to create a full-stack robotics development environment. I wish to unify the design process of kinematics, CAD, control-law, firmware, CAM and simulation into a single skill set applied across the disciplines. That single skill set is programming. Through the perspective of programming all the tasks and relationships of the interconnected disciplines that make up robotics engineering. You can express relationships and interactions as functions and libraries. The purpose for this re-defining of the development process is to speed up robotics development, lower the skill barrier for new developers, and increase the capabilities of the state of the art in robotics engineering.

Bowler Studio is a device manager, scripting engine, CAD package, and simulation tool all in one application. A user can develop the kinematic of an robot arm using the D-H parameters-based automatic kinematics engine. With this kinematics model, the user can then generate the CAD for new unique parts to match the kinematic model. The user can then export the model to an STL, and connect a Bowler 3d printer to Bowler Studio. The printer can print out the part (using the newly generated STL) while the user connects a DyIO and begins testing the servos with the kinematics model. When the print is done, the user can assemble the arm with the tested servos and run the model again to control the arm with Cartesian instructions. Once this is complete, the user can then attach a wiimote to train the robot arm through a set of tasks, recording them with the animation framework built into Bowler Studio. To be sure the arm is moving to the right place, the user can attach a webcam to the end and use Open CV to verify the arm's position, or use the arm (in conjunction with the webcam with Open CV enabled) to track and grab objects (IE "eye-in-hand" tracking).

Care-O-bot

Mobile manipulators are the robots we want, because they're the robots that have the most potential to do the things that we care about: working in our homes and businesses, making things better and faster and easier. Robots have a long way to go before better and faster and easier become a thing that consumers get to experience directly, but with each new and updated platform, we get a little closer. Today, that little bit closer is the new and improved Care-O-bot 4, from Fraunhofer IPA.

The meanwhile fourth generation of this successful development series is more agile and modular than its predecessors and offers various ways of interaction. It also stands out through the use of cost-reducing construction principles. In this way, large parts of its internal construction feature folding sheet metal, which is economical to produce in small quantities. The enhanced agility of Care-O-bot 4 is owed to the patented spherical joints around discreet pivot points on its neck and hips. They extend the robot's working space and allow 360 degree rotations of head and torso.

Based on its modular system design, the fields of application for Care-O-bot 4 are wide-ranging. The robot can be equipped with one, two or no arms at all. The patented spherical joints on the robot's neck and hips are optional. If the intended purpose is to serve drinks, one hand can be replaced by a tray, or the mobile base platform can be used on its own as a serving trolley. Individual robot platforms can be configured for a wide range of applications: a mobile information center in museums, DIY stores and airports, for collection and delivery services in homes and offices, for security applications or as museum robots at attractions –Care-O-bot 4 is a safe and handy human helper at all times.

Previous investigations have shown that social ways of interaction are essential for the acceptance of interactive service robots. As with previous generations, social role models were used as a guiding vision in developing the design and functionality. Using the display integrated in its head and based on the current situation, Care-O-bot 4 is able to display different atmospheres. While the concept for the Care-O-bot 3 was a more reserved, cautious butler, its successor is as courteous, friendly, and affable as a gentleman.

Holonomic System

A robot that uses omni-wheels can move in any direction, at any angle, without rotating beforehand. Such robots called holonomic robots can spin while translating forward at the same heading. A holonomic robot is one which is able to move instantaneously in any direction in the space of its degrees of freedom.

So how does a robot that can move omni-directionally work? The trick is that it uses special omni wheels as shown in figure. The omni-wheel is not actually just a single wheel, but many wheels in one. There is the big core wheel, and along the peripheral there are many additional small wheels that have the axis perpendicular to the axis of the core wheel. The core wheel can rotate around its axis like any normal wheel, but now that there are additional wheels perpendicular to it, the core wheel can also move parallel to its own axis.

Omni Wheel Used in Holonomic Robot

Holonomic and Non-Holonomic System

There are only two types of mobile robots, holonomic robots and non-holonomic robots. If the controllable degree of freedom is less than the total degrees of freedom, then it is known as non-Holonomic drive. A car has three degrees of freedom; i.e. its position in two axes and its orientation. However, there are only two controllable degrees of freedom which are acceleration (or braking) and turning angle of steering wheel. This makes it difficult for the driver to turn the car in any direction (unless the car skids or slides).Non-holonomic robots cannot instantaneously move in any direction, such as a car. This type of robot has to perform a set of motions to change heading. For example, if you want your car to move sideways, you must perform a complex 'parallel parking' motion. For the car to turn, you must rotate the wheels and drive forward.

A holonomic robot however can instantaneously move in any direction. It does not need to do any complex motions to achieve a particular heading. This type of robot would have 2 degrees of freedom in that it can move in both the X and Y plane freely. For e.g., the 3-wheel omnidirectional robot has three differential degrees of freedom. Such a robot has no kinematic constraints, it is able to independently set all the three pose variables: x, y, θ. Holonomic refers to the relationship between controllable and total degrees of freedom of a robot. If the controllable degree of freedom is equal to total degrees of freedom, then the robot is said to be Holonomic.

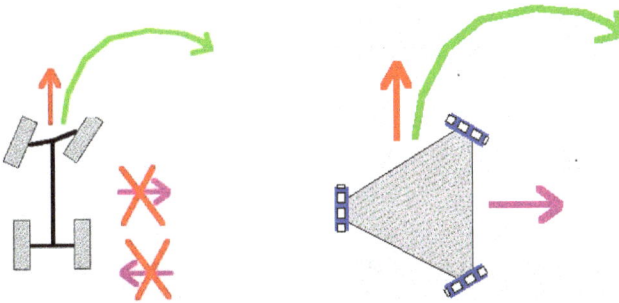

Fig Holonomic and Non-Holonomic Motion Comparision

Wheel Omnidirectional Robot Kinematics

The way in which the individual parts of a robot can move with respect to each other and the environment is called the kinematics of the robot. The forward kinematics specifies which direction the robot will drive in (linear velocities $\dot{x}$ and $\dot{y}$) and what its rotational velocity $\dot{\theta}$ is based on the given individual wheel velocities. The inverse kinematics is a matrix formula that specifies the required individual wheel speeds for given desired linear and angular velocity $(\dot{x}, \dot{y}, \dot{\theta})$ and can be derived by inverting the matrix of the forward kinematics.

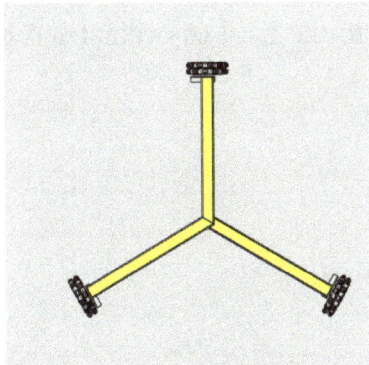

Fig. Holonomic (3-Wheel Omnidirectional) Robot Mechanical Design

Holonomic robot design consisting of three Swedish 90° omnidirectional wheels placed 120 degrees apart is shown in Figure.

Figure shows the geometric diagram of holonomic robot used to find its kinematic model. In the figure, O is the robot center of mass, Po is defined as the vector connecting O to the origin and Di is the drive direction vector of each wheel. The unitary rotation matrix $R(\theta)$ is defined as:

$$R(\theta) = \begin{pmatrix} \cos\theta & -\sin\theta \\ \sin\theta & \cos\theta \end{pmatrix}$$

Where θ is rotation angle in counter clockwise direction.

P_o denotes the position of center of mass with respect to the world frame i.e. $P_o = [x\ y]^T$. The position $[x_i\ y_i]^T$ of each wheel can be given with respect to the center of mass of the robot, i.e., for i=1, 2, 3.

$$P_{oi} = \begin{bmatrix} x_i \\ y_i \end{bmatrix} = R(\theta)L\begin{bmatrix} 0 \\ 1 \end{bmatrix}$$

where L is the distance of wheels from the robot center of mass (O).

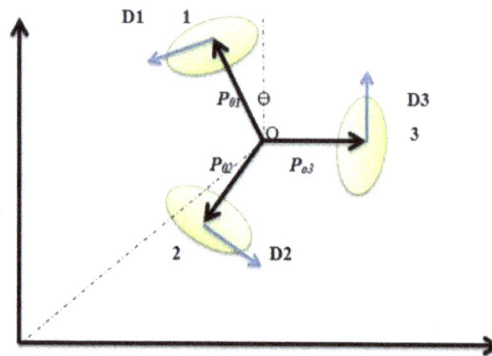

Kinematic Analysis of Holonomic Robot

P_{o1}, P_{o2} and P_{o3} with respect to the local co-ordinates centered at the robot center of mass are given as:

$$P_{o1} = L\begin{bmatrix} 0 \\ 1 \end{bmatrix}$$

$$P_{o2} = R\left(\frac{2\pi}{3}\right) * P_{o1} = \frac{L}{2}\begin{bmatrix} -1 \\ -\sqrt{3} \end{bmatrix}$$

$$P_{o3} = R\left(\frac{4\pi}{3}\right) * P_{o1} = \frac{L}{2}\begin{bmatrix} \sqrt{3} \\ -1 \end{bmatrix}$$

The drive directions D_i of each wheel can be obtained by

$$D_i = \frac{1}{L} R\left(\frac{\pi}{2}\right) P_{oi}, i = 1, 2, 3$$

which yields

$$D_1 = \begin{bmatrix} -1 \\ 0 \end{bmatrix}$$

$$D_2 = \frac{1}{2}\begin{bmatrix} \sqrt{3} \\ -1 \end{bmatrix}$$

$$D_3 = \frac{1}{2}\begin{bmatrix} 1 \\ \sqrt{3} \end{bmatrix}$$

The position and velocity of each wheel with respect to the world frame are then expressed by, for $i = 1, 2, 3$

$$R_i = P_o + R\left(\theta + \frac{2\pi}{3}(i-1)\right)P_{oi}$$

$$v_i = \dot{P}_o + \dot{R}\left(\theta + \frac{2\pi}{3}(i-1)\right)P_{oi}$$

Translational velocity of each wheel is obtained as:

$$V_i = v_i^T\left(R\left(\theta + \frac{2\pi}{3}(i-1)\right)D_i\right)$$

This gives

$$V_1 = -\cos(\theta)\dot{x} - \sin(\theta)\dot{y} + L\dot{\theta}$$

$$V_2 = \cos\left(\frac{\pi}{3} - \theta\right)\dot{x} - \sin\left(\frac{\pi}{3} - \theta\right)\dot{y} + L\dot{\theta}$$

$$V_3 = \cos\left(\frac{\pi}{3} + \theta\right)\dot{x} - \sin\left(\frac{\pi}{3} + \theta\right)\dot{y} + L\dot{\theta}$$

which can be written in matrix form as:

$$\begin{bmatrix} V_1 \\ V_2 \\ V_3 \end{bmatrix} = P(\theta)\begin{bmatrix} \dot{x} \\ \dot{y} \\ \dot{\theta} \end{bmatrix}$$

Where,

$$P(\theta) = \begin{bmatrix} -\cos(\theta) & \sin(\theta) & L \\ \cos(\theta)\left(\dfrac{\pi}{3}-\theta\right) & \sin\left(\dfrac{\pi}{3}-\theta\right) & L \\ \cos(\theta)\left(\dfrac{\pi}{3}+\theta\right) & \sin\left(\dfrac{\pi}{3}+\theta\right) & L \end{bmatrix}$$

is the inverse kinematics matrix. Matrix $P(\theta)$ is always singular for any value of θ. Thus,

$$P^{-1}(\theta) = \begin{bmatrix} -\dfrac{2}{3}\cos(\theta) & \dfrac{2}{3}\cos\left(\dfrac{\pi}{3}-\theta\right) & \dfrac{2}{3}\cos\left(\dfrac{\pi}{3}+\theta\right) \\ \dfrac{-2}{3}\sin(\theta) & \dfrac{-2}{3}\sin\left(\dfrac{\pi}{3}-\theta\right) & \dfrac{2}{3}\sin\left(\dfrac{\pi}{3}+\theta\right) \\ \dfrac{1}{3L} & \dfrac{1}{3L} & \dfrac{1}{3L} \end{bmatrix}$$

is the forward kinematics matrix which satisfies

$$\begin{bmatrix} \dot{x} \\ \dot{y} \\ \dot{\theta} \end{bmatrix} = P^{-1}(\theta) \begin{bmatrix} V_1 \\ V_2 \\ V_3 \end{bmatrix}$$

Perceptual Control Theory

Perceptual Control Theory (PCT) is a revolutionary and exciting theory of human behavior, one that invalidates much of what currently passes as psychological research and existing theories of human behavior.

Concieved of by William Treval Powers in 1952 and developed by him across more than 60 years with tireless brilliance, supported by followers loosely organized as the Control Systems Group.

Essentially, PCT views people as purposeful, living control systems, whose behavior shapes its consequences instead of the other way around. PCT is a feedback-governed view of human behavior. It holds that we target certain variables for control and we compare our perceptions of the current state of those variables with our goal state or reference condition for those variables. If unacceptable gaps exist, we behave in ways that serve to close those gaps. Thus it is that our behavior serves to control our perceptions.

There are, however, other actors and factors at work that influence the same variables we are trying to control. Ordinarily these disturbances as they are known in PCT are compensated for and pose no problem. On occasion they can prove overwhelming. Our control is far from perfect.

PCT abounds with insights, implications and new directions for researchers, those who simply want to understand human behavior and those who would manage human performance in the workplace.

PCT lends itself well to the fields of artificial intelligence and robotics because it provides an exact mathematical framework to model psychological processes.

Two good examples of PCT would be Richard Kennaway's six-legged robot called Archy and Robby the Robot by Walter Fritz". Another fascinating development has been to use PCT to model human motor control. These models have been described from a theoretical basis (e.g. Powers, 1999). Recently, these have been realized as machines. For example, Adam Matic's robot arm that uses visual and pressure sensors". In complementary work, Adam Matic has modelled extensor and flexor muscle action using PCT.

Robot models built using PCT are very simple in design and include few components compared to some of the complex models that try to model specific behaviors rather than using negative feedback to control perception. This suggests that PCT is parsimonious – and may be hitting on what systems are used in nature. For example, in order to grasp an object effectively, it is not necessary for a robot to perceive the shape of the object, as modelled by Rodrigues.

Programmable Matter

Programmable matter is a material than can be programmed to change its shape, color, density, and so on, either autonomously or in response to user input. As it includes modular robots and intelligent "paper" that folds into different shapes, programmable matter does not necessarily need to be on a nanoscale. Nanoscale applications may include swarms of nanorobots called "claytronics", synthetic cells that perform novel functions, or even quantum wells that behave similarly to actual atoms.

Let's take a look at some of the current research into programmable matter, and where it could eventually end up. We'll begin by discussing two different engineering approaches to realizing programmable matter—modular robotics and metamaterials—and then consider how they might ultimately converge.

Modular Robotics

The modular robotics approach to programmable matter aims to develop robotic units capable of arranging themselves into arbitrary configurations. For example, in 2013, a team of MIT engineers developed the first prototypes of what they called M-Blocks, tiny cube-shaped robots capable of propelling themselves without any external moving parts. Grouped together, the M-Blocks could organize themselves into simple cube-based configurations.

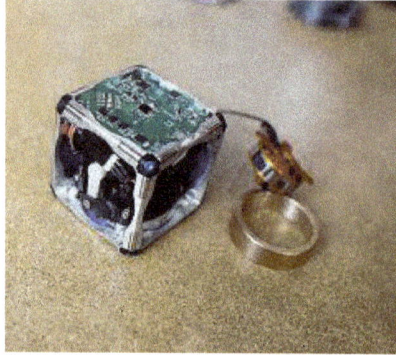

An M-Block with its internal flywheel pulled out for display. Each M-Block measures 50 cubic millimeters, with a total mass of 143 g.

To locomote, M-Blocks pivot around their edges using the principles of angular momentum. Inside each robot is a flywheel that spins at up to 20,000 rpm. When a sudden braking is applied to the flywheel, it transfers its momentum to the M-Block, propelling it forward.

To arrange themselves in groups, each M-Block is equipped with two cylindrical magnets embedded along each edge, which keep the M-Blocks in place as they pivot over other M-Blocks. An additional eight magnets on each face help keep the M-Blocks in a square alignment.

Not all modular robots are cube-shaped. Another team from MIT took inspiration from the geometrical complexity of proteins to develop moteins (motorized proteins), chain-like assemblies composed of simple robotic modules. Moteins are based on a paper from 2011 that outlines a technique whereby any 2D or 3D shape can be achieved by folding strings made up of simple robotic subunits.

A motein chain composed of four one-centimeter modules.

"It's effectively a one-dimensional robot that can be made in a continuous strip, without conventionally moving parts, and then folded into arbitrary shapes.

Real-time Path Planning

Path planning of mobile robots is one of the key issues in robotics research on the problem of a robot finding a collision-free path from beginning to goal in the presence of obstacles. Depending on the environment surrounding the robot, it can be classified as follows:

(1) Path planning for static obstacles in completely known environment.

(2) Path planning for static obstacles in unknown or partially known environment.

(3) Path planning for dynamic obstacles in completely known environment.

(4) Path planning for dynamic obstacles in unknown or partially known environment.

To the former two cases, that is, path planning in the presence of static obstacles, there are various approaches, such as c-space method and artificial potential field method. They may obtain perfect results under some certain condition. But in fact, robots mostly work in dynamic uncertain environment that includes static obstacles with unknown position and dynamic obstacles with uncertain trajectory. So the collision-free navigation of mobile robots in dynamic uncertain environment is so complex that it is still an intractable topic by far.

Many real-time navigation systems have been developed recently. Fujumura proposed an algorithm for obstacles with known trajectories. Tang treated instantaneous dynamic obstacle as static and proposed an optimal path based on a grid method called dynamic grid method. Fiorinio and Shiller advanced a view of velocity obstacle that plans on-line motion based on relative velocity between the robot and moving obstacles. Rude used Time-Space coordinates to translate robot's path planning in 2D dynamic environment into static path planning in 3D space. Takeshi used three-tier fuzzy control method to adjust robot's motion direction and velocity for detouring dynamic obstacles.

It is difficult to give obstacle's reliable motion information obviously, e.g. velocity and direction. In fact, such real-time information is hidden in next-time obstacle's position. So we can obtain obstacle's motion information provided that next-time the obstacle's position is predicted by some prediction algorithm. Some navigation systems combine path planning with obstacle motion prediction. A prediction model takes object's previous information and uses various methods to predict the motion trend of the object to get its next position. In motion prediction, numerical prediction approaches, such as curve fitting or regression methods (Sen and Srivastava), are widely used because they are simple and convenient. Other models that may be used to improve the prediction results are hidden Markov stochastic models (Zhu), the Grey prediction (Luo and Chen), etc. The Kalman Filter (Kalman) is also used for predicting future positions and orientations of a moving object in dynamic environments. The Kalman Filter's recursive nature enables it estimate a real-time system state.

The motion trajectories of moving obstacles are predicted by using an autoregressive (AR) model in this paper. Compared with other models above, the AR model is so simple and its algorithm is so fast that the robot can give a quick response when encountering obstacles. In the meanwhile, prediction precision can also be guaranteed. Positions of moving obstacles are sampled by the robot's sensor system. The sampling position information on dynamic obstacles is treated as instantaneously static. With current sampling positions, the AR model predicts future obstacle's positions in the next sampling duration. Then the robot's motion path is planned with such predicted positions. Thus, dynamic collision-free path planning is translated into static one. Combining global planning with local planning, we propose an on-line real-time path planning technique based on polar coordinates in which the desirable direction angle is taken into consideration as an optimization index. Detecting unknown obstacles with local feedback information by the robot's sensor system, this approach orients the desirable direction of the mobile robot so as to generate local sub-goal in every planning window. As a result, the difference between real direction angle and desirable direction angle of robot motion steers the mobile robot to detour collisions and advance toward the target without stopping to re-plan a path when new sensor data become available.

Robot Interaction Language

The number of robots in our society is increasing rapidly. An easy way to communicate with service robots, such as Roomba or Nao, is natural speech. Prasad et al. even go as far as describing speech interaction with robots as a"Holy Grail". However, the limitations in speech recognition technology for natural language is a major obstacle for the general introduction of speech interaction for robots. At times, current speech recognition technology is not good enough for it to be deployed in natural environments where the ambience influences its performance. One of the problems that speech recognition technology is facing is the inherent properties of natural language. Examples are dialects, ambiguity in context, grammar irregularities, and homophones (words that sound the same but have different meanings). As a consequence, miscommunication occurs frequently between the user and the robot, which leads to a considerable frustration for the user. The Palm Company faced a similar problem with handwriting recognition for their handheld computers in the 1990s. To overcome the insufficient recognition accuracy, they invented an artificial alphabet: Graffiti. It is easy for users to learn and easy for the computer to recognize this new alphabet. The Robot Interaction Language (ROILA) takes a similar approach by offering a speech recognition friendly artificial language that is easy for users to learn and easy for machines to understand.

Normal alphanumeric gestures

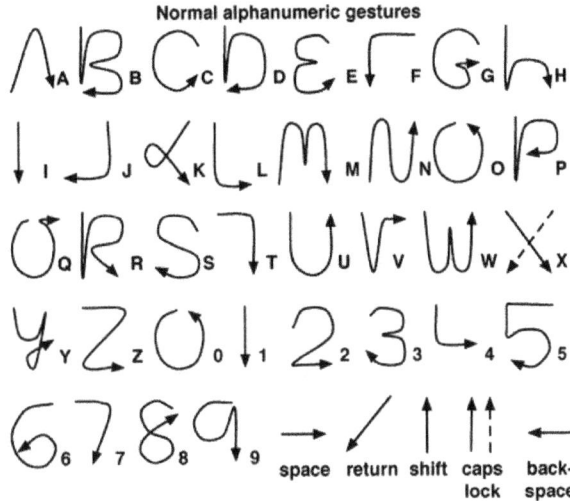

Graffiti from Palm Inc.

An artificial language is a language that is deliberately invented or constructed, especially as a means of communication in computing or information technology. However, within the domain of artificial languages, both spoken and nonverbal languages such as formal languages and programming languages exist. Several artificial languages exist, and Esperanto might be one of the most acknowledged. To the best of our knowledge, none of the available spoken artificial languages (such as those mentioned in5) were optimized for human machine interaction but were rather created to facilitate human-human communication ("also known as universal languages"). Therefore, the focus of our research was to design a speech-recognition friendly artificial language that humans could use to talk to robots. First attempts at creating a speech-recognition friendly language have been made by constraining the use of a natural language. Rosenfeld, Olsen, and Rudnicky argued that constraining language is a plausible method of improving recognition accuracy. The user experience of an artificially constrained language was evaluated within a movie-information dialog interface, and it was concluded that 74% of the users found the constrained language interface to be more satisfactory than a natural language interface. Another example is the constrained command language proposed in that is intended to be used for interacting with various appliances.

Several attempts have been made to ensure efficient speech interactions between robots and human beings. Besides efforts in improving verbal human–robot interactions, spoken languages for robot–robot communication have started emerging,1 where the focus can also be to teach robots languages along the way of improving human–robot interaction in the long run.11 Still, the main thrust of research in human–robot verbal interaction is focusing on providing controlled command languages to interact with robots, and first results have become available.12,13 These constrained languages are inherited from natural languages, and, are, therefore, potentially easy to learn. In addition, constrained languages may limit the vocabulary, grammatical structures, or even go as far as modifying words.

Another stream of research that improves speech recognition is the use of a multimodal input, for example, by using cameras within the robots' environment to record the gestures of the user.14–16 Therefore, the robot is not reliant on only one modality, and, consequently, there is a prospect of better human–robot communication. As reported in,17 the robot tracks the gaze of the user when the object or the verb of a sentence in a dialogue is undefined or ambiguous.

However, little is known about how difficult it is to learn such constrained languages, and the available data on the efficiency of such afore-mentioned multimodal speech systems are still scarce. For example, in systems such as those described in,18,19 speech recognition technology has been shown to work with robots in controlled settings, but their tests were only carried out with a limited number of speakers (in both cases, 3 speakers). Therefore, it is debatable as to whether such systems would work outside the lab and when exposed to a large group of users.

Given the high potential for speech interaction between users and robots, we explore a spoken artificial language that aims at achieving a new balance between, on the one hand, being easy to learn for users and, on the other hand, being easy to recognize for robots. Humans are incredibly adaptive, and an artificial language that utilizes this advantage might have the potential to outperform natural languages (in terms of recognition accuracy) in the interaction between users and robots. We, therefore, go beyond constrained languages by developing and evaluating a ROILA. This truly new artificial language is optimized for automatic speech recognition and, in theory, requires minimal learning effort from humans. There have been first efforts in designing speech-recognition friendly artificial languages to talk to machines, but the results of these studies are fairly limited. In both studies mentioned, the vocabulary size was small, and the languages had no grammar, that is, commands consisted of solitary words. Moreover, no results of formal evaluations with users are available.

It is a clear disadvantage that users need to learn ROILA to interact with robots. However, similar to learning to type with ten fingers, it might pay off in the long run. We, therefore, need to keep the efficiency of ROILA in mind by monitoring the learning effort and the potential gain in recognition accuracy.

References

- Robotic-systems, bin-picking: teqram.com, Retrieved 11 July 2018
- The-Promise-and-Peril-of-Programmable-Matter: engineering.com, Retrieved 27 April 2018
- Structured-vs-random-bin-picking: motioncontrolsrobotics.com, Retrieved 10 May 2018
- Bowlerstudio-a-robotics-development-platform-6423: hackaday.io, Retrieved 22 June 2018
- Kinematic-analysis-of-holonomic-robot: bharat-robotics.github.io, Retrieved 24 April 2018
- Programmable matter: nanosupermarket.org, Retrieved 14 March 2018

Permissions

Index

www.ingramcontent.com/pod-product-compliance
Lightning Source LLC
Chambersburg PA
CBHW062004190326
41458CB00009B/2967